T0249674

Components and Mechanisms

Evolutionary science deserves to be much better understood.

Lynn Margulis

Seen in the light of evolution, biology is, perhaps, intellectually the most satisfying and inspiring science. Without that light it becomes a pile of sundry facts—some of them interesting or curious but making no meaningful picture as a whole.

Theodosius Dobzhansky

The evolutionary process of natural selection described by Darwin is just one of the mechanisms leading to the elaboration of a new species.

Alexandre Meinesz

Evolution
Components and Mechanisms

David Zeigler, PhD
Department of Biology
University of North Carolina at Pembroke
NC, USA

ELSEVIER

AMSTERDAM • BOSTON • HEIDELBERG
• LONDON • NEW YORK • OXFORD •
PARIS • SAN DIEGO • SAN FRANCISCO •
SINGAPORE • SYDNEY • TOKYO

Academic Press is an imprint of Elsevier

Academic Press is an imprint of Elsevier
32 Jamestown Road, London NW1 7BY, UK
225 Wyman Street, Waltham, MA 02451, USA
525 B Street, Suite 1800, San Diego, CA 92101-4495, USA

Copyright © 2014 Elsevier Inc. All rights reserved.

No part of this publication may be reproduced, stored in a retrieval system or transmitted in any form or by any means electronic, mechanical, photocopying, recording or otherwise without the prior written permission of the publisher

Permissions may be sought directly from Elsevier's Science & Technology Rights Department in Oxford, UK: phone (+44) (0) 1865 843830; fax (+44) (0) 1865 853333; email: permissions@elsevier.com. Alternatively, visit the Science and Technology Books website at www.elsevierdirect.com/rights for further information

Notices
No responsibility is assumed by the publisher for any injury and/or damage to persons or property as a matter of products liability, negligence or otherwise, or from any use or operation of any methods, products, instructions or ideas contained in the material herein.

Because of rapid advances in the medical sciences, in particular, independent verification of diagnoses and drug dosages should be made

British Library Cataloguing-in-Publication Data
A catalogue record for this book is available from the British Library

Library of Congress Cataloging-in-Publication Data
A catalog record for this book is available from the Library of Congress

ISBN: 978-0-12-800348-0

Cover Image: From Shutterstock, bird fossil of unknown species, Green River, Wyoming, USA.

For information on all Academic Press publications
visit our website at elsevierdirect.com

Typeset by TNQ Books and Journals
www.tnq.co.in

Working together
to grow libraries in
developing countries

ELSEVIER   Book Aid International

www.elsevier.com • www.bookaid.org

Contents

Preface   vii
Acknowledgements   xiii
Biography   xv

1. Evolution   1

2. Natural Selection   9

3. Adaptation   23

4. Competition   31

5. Genetics Basics & Mutations   39
   Mutations   42
      Point Mutations   43
      Chromosomal Mutations   45
      Mutations from Transposable Elements   48
      Epigenetics   49

6. Transposable Elements, Viruses, and Genomes   55

7. Horizontal Gene Transfer   61

8. Neutral Evolution   69
   The Redundant Genetic Code   70
   Neutral Nonsilent Mutations   72
   Mutations in Introns   74
   Mutations in Other Noncoding DNA   75
   Some Chromosomal Mutations Can Be Neutral   76

9. Genetic Drift   79

10. Environment   85

11. Development   93

12. Symbiosis   101

13. Speciation   111
   Allopatric Speciation   112
   Sympatric Speciation   116
   Parapatric Speciation   118
   Speciation by Polyploidy   119
   Speciation without Cladogenesis   120

14. Micro- and Macroevolution   123

15. Homology   129

16. Imperfection   137

17. The Fossil Record and the History of Life   145

18. Contingency and Evolution   151

19. Opportunity   159

20. Phylogeny—The Tree of Life   165

21. Progress—Purpose?   171

References   177
Appendix 1   179
Appendix 2   185
Index   189

Currently, the science of evolution emerges as a new synthesis waiting for the synthesizers.

Antonio Fontdevila

Science is a glorious enterprise.

Niles Eldredge

An obvious question to begin with is: do we really need yet another book on the basics of evolution? In the past two to three decades numerous evolution books have flooded the market—all attempting in one way or another to either explain evolution, or to defend evolution in the ongoing "conflict" between science and religion as it concerns evolutionary theory. This latter area has become something of a cottage industry that seemingly feeds on itself and generates an almost endless stream of books, each attempting some new or different take on this so-called conflict. A few of these books are quite good and well worth reading, but this book will touch only minimally on "the conflict."

Other evolution books make Darwin and the history of evolutionary theory the major focus of discussion. This book only mentions Darwin as needed to explain the modern basics of evolution—of which "Darwinism" is still a key component. Some of the major textbooks in evolution take a track that is heavy on genetics, mathematics, and modeling, and again, this book only touches lightly on these approaches. Yet another group of recent books addresses and explains the abundant evidence which supports evolutionary theory, and since some of these are excellent works this book does not dwell long on their vast riches of supporting evidence. Finally, several excellent and recent books have specifically addressed the new findings and understandings deriving from the flood of evidence coming from recent genomic studies: this book does include some examples and findings from this rich and newly-elucidated store of information—but not, however, in great detail.

This book mainly emphasizes the fact that our understanding of biological evolution is an ongoing synthesis of ideas, discoveries, and understandings that grows daily, thanks to the thousands of scientists who make relevant discoveries and contributions on a regular basis. Only in the last decade or two has genomics made significant contributions to our understanding of evolution. Recent significant fossil finds have also added greatly to our understanding of several major clades and transitions. Relatively new discoveries concerning the area of symbiosis have added support to the claims of Lynn Margulis that this has played a significant role in the evolutionary process. The increased recognition of the huge role that contingency has played in the evolution of life on Earth represents another recent theoretical aspect of the modern consensus. These and other contributing areas of knowledge have made the topic of evolution more complex—one which now requiringes a good bit of background knowledge to feel comfortable with. In short, evolution is not a static discipline—far from it. Organic evolution is a rich and evolving scientific discipline of amazing depth and detail—one that every educated person should at least attempt to appreciate, since it is the ultimate reason why we and some 10,000,000 other living species now inhabit the Earth.

Though it has been said that evolution by natural selection is a simple idea, today the subject uses a tremendous amount of data from a number of specialty areas, and it carries with it many interesting implications. Stephen J. Gould's 2002 *The Structure of Evolutionary Theory* at the 1,343rd page of text illustrates this point well, as do some of Ernst Mayr's larger works. Books of this type were written for those who are already well steeped in the basics of evolutionary thought, while this far smaller book is aimed at a broader audience, including those with only an introductory understanding of biology and evolution.

I hope that this book might serve as a sourcebook for review, as an introduction to evolution for the curious learner, or perhaps as a text in an introductory course in evolution. My somewhat unique framework will involve taking evolution apart and explaining separately, chapter by chapter, most of the essential components and mechanisms which together make up our current understanding of the evolutionary process.

We cannot avoid the fact that every large and encompassing area of biology (or any other science) accumulates over time a body of scientific jargon, which hopefully enhances understanding—and at the same time reduces the verbiage needed to communicate within the discipline. While this is unavoidable and even necessary, it creates a real need for everyone involved to understand and agree on the meaning of these numerous scientific terms.

Though I doubt that any two evolutionists would agree precisely on the most essential terms and components of modern evolutionary theory, I believe that most would accept that the majority of my chapter titles represent many of the important concepts and topics needed for a fundamental understanding of evolution. Additional necessary jargon will appear in the book. I have tried to incorporate an explanation of any additional terms in the most logical places. In short, I have taken something of a "reductionist" approach—attempting to cover and address the many important concepts separately, while realizing that some topics will need to be cross-referenced to others elsewhere in the book. This differs from many of the current evolution books I have encountered, which weave several threads of thought into long chapters. This sometimes work well, but too often the narrative becomes convoluted and strays from any single train of thought—which is especially difficult for the reader lacking a firm background in the subject of evolution.

My hope was to write a book that would be a clear and accessible aid to understanding either parts of, or much of the whole of, evolutionary theory as needed by the reader. I will not claim that this approach is superior to that of other, more experienced writers in this area. I admire and have profited much from many other excellent books on evolution; please do note the suggested reading list at the end of the book (Appendix II) in which I have listed and commented on other relevant books by truly talented scientists and/or writers.

Since Darwin's day, our understanding of evolution has itself evolved significantly and continues to do so rapidly. Though there were some earlier periods of relative stasis, these last two or three decades have certainly seen rapid and numerous advances in our understanding of evolution, and some of these have affected the topics covered in this book. Punctuated equilibrium, clade, and genome are only three of the terms which have recently been introduced, and certainly no one can understand evolution well without being aware of these new ideas.

Our increasing understanding of the record of evolution which can be found in the DNA of living organisms continues to revolutionize our understanding of phylogeny, the timing of change, the significance of some symbiotic events, and the origin of some DNA segments (duplications, lateral transfers, viruses, plasmids, "captured" symbionts, transposable elements, etc.). The genomic DNA of the Earth's species represents the most character-rich area of study for determining the actual relationships between the millions of living species, and the future undoubtedly holds many more significant discoveries and surprises in this area.

The fact that several processes other than natural selection are responsible for the process of evolution is still not fully appreciated by everyone—including even some in the scientific community. Though undoubtedly natural selection is largely responsible for most of the adaptations seen in living organisms, these organisms are not just collections of adaptations—this fact being perhaps most evident at the level of the genome, where waste, excess, and genomic conflict seems to be present. Some of these other contributing processes are discussed in the book, since these concepts are now essential to our understanding of organic evolution.

Almost certainly the most repeated quote concerning evolution is the one penned by Theodosius Dobzhansky: "Nothing in Biology makes sense except in the light of Evolution." Even some who fully accept evolution have said that this is a bit of an overstatement, but at the level of ultimate explanation, all biological phenomena arose through the evolutionary process, and evolution does have some relevance to almost every issue in biology. As some have said concerning natural selection: never has so simple an idea explained so much of the world. To quote another legendary evolutionary writer: "I do not think that evolution is supremely important because it is my specialty. On the contrary, it is my specialty because I think it is supremely important." (George Gaylord Simpson).

Anyone who approaches a full understanding of evolution will realize just how all-encompassing evolutionary theory is, and how it does in fact connect to all things biological. On a personal note I must say that I feel both fortunate and privileged to have spent so many years learning and engaging in my own small way with such a grand and important body of work as the one that now explains and supports evolutionary theory. Engagement with a topic of this caliber and magnitude is truly what "the life of the mind" is all about.

Of the many great scientific breakthroughs of the last two centuries, many feel that evolutionary theory, from the ideas of Darwin, to the evolutionary understanding of genomes and selfish genes, is the greatest set of scientific insights ever to be perceived by humans. It has certainly revolutionized our view of life on Earth, including of course, that of *Homo sapiens*—our own species. It is my hope that this small contribution will aid in making more of us literate in this immensely important and interesting domain of knowledge.

David Zeigler
Department of Biology, University of North Carolina at Pembroke

Acknowledgements

The author would like to sincerely thank Pat Gonzalez and Kristi Gomez (editors), as well as Karen East and Kristy Halterman (project managers), who were a pleasure to work with during the process of bringing this book to completion. Thanks to all those at Elsevier and Academic Press who were involved in this project.

I also want to thank all scientists everywhere who have increased our objective and rational understanding of our universe, our world, and the life on this planet—including ourselves. Most of my heroes are scientists, and I try to honor their work in the classroom and in my writing. To mention just two from evolutionary biology: thanks to Ernst Mayr and Richard Dawkins who's books I have read and continue to read for insights into the workings of the living world. What riches of understanding workers of this caliber have bestowed on us!

Biography

David Zeigler is currently a professor of Biology at the University of North Carolina at Pembroke where he teaches evolution, animal behavior, parasitology, and invertebrate zoology. His graduate degrees are from the University of North Texas in Denton, TX where he worked on insect behavior. He is also author of the book *Understanding Biodiversity* (2007) published by Praeger.

Chapter 1

Evolution

The reality is that all sorts of things evolve: galaxies, stars, a person's thinking, a government's policies. But none of these qualify as evolution as biologists understand the term.

David Barash

If evolution meant only gradual genetic change within a species, we'd have only one species today—a single, highly evolved descendant of the first species.

Jerry A. Coyne

That extant living forms are derived from earlier, often simpler forms, and that all living forms are related by descent, is as much an established fact as anything in science can ever be.

John Dupre

It would seem advisable to start this book by defining the term under discussion—evolution. In practice, some terms are more difficult to define than others—and for some the best definition does not go far toward making the term familiar to the reader. A good example would be the simple word "animal". Almost everyone alive has used this word repeatedly in their lives, though surely not even one in a hundred people could give a proper scientific definition of what that term means. Including all the pertinent terms, an animal is a eukaryotic multicellular organism lacking cell walls that develops from an embryo and utilizes holozoic heterotrophic nutrition. See what I mean? And even this definition has a few problems and exceptions. The term evolution is not quite that hard to define, and can in its broadest meaning

Evolution. http://dx.doi.org/10.1016/B978-0-12-800348-0.00001-8
Copyright © 2014 Elsevier Inc. All rights reserved.

refer simply to change in some entity or group, and in this vein can rightly be applied to the following:

- The evolution of the swimsuit
- The evolution of the automobile
- The evolution of the universe
- The evolution of Hemingway's writing style
- The evolution of the Republican Party
- The evolution of Darwin's thinking on evolution

The change referred to in such broad usage is change that can, at least in theory, be followed and compared from one point in time to another. Some definitions of evolution include the idea of progression—as would be generally obvious in the evolution of the automobile and perhaps in Hemingway's writing style, but a modern understanding of biological evolution is not tied to the concept of progression (Chapter 21). One problem with progress is that not all biologists and evolutionists agree on the meaning of that term. Another problem is that progress implies that a later evolutionary stage is in some way better or superior to an earlier stage—which is a value judgment—which is not what science involves itself in (judging whether something is better or worse than something else). Be warned that anytime we venture into judgments of value, we venture beyond the boundaries of science.

The term evolution is also not generally applied to either change involving simple aging or change resulting from the processes of decay and decline. Therefore it would not be appropriate usage to say either that an old man evolved from a child, or that a rusty old automobile evolved over the years from a new showroom model. Diseases like skin cancer or smallpox would most likely (if survived) result in some physical changes to an individual that likewise would not appropriately be termed evolution, and in biological usage, individuals do not evolve anyway—only populations of organisms evolve. The evolving Universe example earlier includes aging, but the resulting events and processes hardly look (currently) like widespread decay and decline since new stars and planets are still forming, along with possibly life and other complex processes of chemistry, geology, etc.

The more obvious and slightly narrower meaning of evolution refers to "organic evolution", or the evolution of life on Earth. A definition for this meaning (which is the topic of this book) needs to be more narrow and to contain more detail than saying simply change—or even Darwin's "descent with modification". For the past few decades, biologists have usually defined evolution as change in the genetic makeup of populations. This means that over many generations, the frequencies of genes and alleles are potentially changing within populations (alleles are defined as variations of a gene). Many genes have two or more alleles, while some have only one form and therefore technically have no alleles. We would say such genes are *fixed* within the population. Change in genetic makeup also includes the addition and deletion of genes. Many of the mechanisms responsible for genetic change are discussed in more detail in Chapters 5 through 9.

A term now being commonly used is "genome", which includes not only the genes that code for detectable traits, but also all the DNA of a species—whether it is coding, or whether it fits under the once common title of "junk DNA". Books and articles now appear frequently which refer to the evolution of the genome, and some of these works give convincing evidence that the non-coding DNA is also evolving (often quite rapidly), as well as having many important effects on the coding genes. And it is not just changes in the DNA bases or the population gene and allele frequencies that count as genomic change, it also includes rearrangements of genome "architecture" that occur as the result of chromosomal mutations such as inversions and translocations. Therefore, it is probably now best to go ahead and speak of evolution as the change in population *genomes* over time, and Chapters 5 and 6 give further support for this point. One additional point to make is that for eukaryotic organisms, there may be two or more genomes present in cells. The main genome is of course that contained in the chromosomes within the nucleus—the nuclear genome, but mitochondria too contain a small remnant of their original genome descended from their bacterial ancestors (explained further in Chapter 12). In photosynthetic eukaryotes, the chloroplasts also contain a separate genome of a few hundred genes descended from their cyanobacterial ancestors (also covered in Chapter 12). Change in any of these inherited genomes constitutes evolutionary change, but simply stating genomic change should suffice to cover all inherited genomic content.

In addition to his idea of descent with modification by means of natural selection, Darwin also argued successfully the case of common descent—the idea that all species can trace their ancestry back to one or a very few original life forms. Previously, many evolutionists like Lamarck argued that all evolution occurred in the form of "anagenesis". Anagenesis (also called phyletic evolution) refers to evolution within a single line of descent where speciation (or cladogenesis) does not occur. In short, A evolves into B, which evolves into C, which evolves into D, with no splitting of the one ancestral line. Though such anagenetic change probably has occurred in many lines over shorter periods of geologic time, the estimated 10,000,000 or more species alive today (and the even greater number of extinct species) arose mainly through untold millions of cladogenetic speciation events over the long history of life on this planet. Speciation or cladogenesis is the primary answer to why biodiversity is so diverse, and many books covering the evolution of specific groups of organisms are filled with cladograms illustrating these phylogenetic patterns of common descent from common ancestors. Phylogeny is defined as the evolutionary history of a group of species (or groups of species at higher taxon levels), and phylogenies are typically illustrated through figures of these branching cladograms. Chapters 13 and 20 cover more ground on these important topics.

It is interesting, and I believe problematic, that the majority of definitions of evolution fail to touch on common descent—Darwin's other major contribution to evolution-ary understanding (with natural selection being his primary contribution). Had Lamarck known about genes, he would likely have had no problem with most modern definitions that stop with something like: "changes over time in the genetic makeup of populations" (though his thinking on mechanisms would have still been wrong). That being the case, it seems that a modern definition of evolution really should incorporate Darwin's important contribution of common descent. A possible definition that would do so might read something like: *Organic Evolution is change in the genomes of populations and species over multiple generations and the cladistic diversification of populations and species over geologic time that regularly results from some of these genomic changes.*

This slightly longer definition still does not speak to mechanisms, but that is unnecessary in a definition, just as a definition of an airplane would not speak to the mechanics and physics of lift and thrust. At least this expanded defini-tion broadens the scope to include the whole genome—and also includes the concept of cladistic divergence (common descent), which is fundamental to our modern understand-ing of evolution.

Two other examples of evolution that have many par-allels (yet some differences) with organic evolution are the evolution of human cultures and languages. Richard Dawkins gives an excellent discussion of cultural evolution in his book *The Selfish Gene,* and several books have been written addressing the evolution of languages. Researchers have even worked out a phylogeny of languages, complete with cladograms illustrating the relatedness and groupings of languages. English is, for example, more closely related to German than it is to Russian. However interesting these topics are, they are not the topic of this book, which deals with organic evolution—the evolution of biodiversity on planet Earth.

One additional topic that appears in some books deal-ing with evolution is the origin of life question. Some (including this author) view this topic as something separate from evolution, though it is certainly a scientific question where numerous researchers have made progress in addressing the possible chemical pathways from nonlife to life. Additionally, this is a subject for which we will likely never have a single confident answer, though researchers may eventually narrow down the possibilities to a very few alternatives.

I would like to end this first chapter by addressing briefly the often quoted phrase that "evolution is a *just a theory*". Of course, this phrasing is most often voiced by those who reject or have problems with the concept of evolution—especially as it applies to humans—and mostly for religious reasons. Many have explained repeatedly that in science the term theory does not mean the same thing as hypothesis, as it does in common usage. In science, the term theory is reserved for an important concept of broad explanatory power that has been shown through significant research and evidence to be the strongest and best-supported concept available. In short, the concept is important and very well supported by objective evidence. Evolutionary theory now stands as an extremely well-supported concept that explains an amazing amount of what we see in the living world. Because there is no scientific contender that explains the living world anywhere nearly as well, and because there is no empirical evidence that puts the modern concept of biological evolution in doubt, many scientists like Richard Dawkins have taken to calling evolution a fact.

If indeed facts are taken to exist (and I believe they do), evolution certainly looks for all the world like a fact. Science usually claims not to be in the business of discovering ultimate truth or facts. Scientific findings are typically said to be provisional—until a better or more accurate description or explanation comes along. This is a safe and humble stance that prevents scientists from looking bad when some accepted idea is overturned—as has occurred occasionally throughout the history of science. Lamarck's explanation of the nature and mechanisms of evolution was wrong, so his accepted scientific explanation was overturned by Darwin's basic explanation, which is still going strong—though with many later additions and modifications.

What scientists are actually saying when they speak of the provisional status of scientific knowledge is that they are going to remain open-minded. Scientists often claim that the creationists are closed-minded (which in fact they are on the topic of evolution), and they would never want this description turned back on them by any group—therefore the official open-minded stance of "provisional knowledge" reigns in the scientific community, and again, some scientific "knowledge" is clearly recognized by scientists to be still tentative. However, most scientists believe there is no chance whatever that new data might someday overturn our understanding that nucleic acids contain the genetic information of living organisms, that the Earth is the third planet from the Sun, or that organic evolution has occurred. Most scientists would have as much confidence in these scientific conclusions as in any other fact they would admit to accepting—such as the fact that copper is softer than steel.

If convincing evidence was forthcoming that our understanding of evolution was basically flawed, scientists should be willing to be open-minded and look at this evidence, though it undoubtedly would be viewed initially as preposterous—due to the current abundance of evidence that supports evolution. The theory of the endosymbiotic origin of eukaryotic cells was judged to be preposterous by many scientists for a while, but Dr. Lynn Margulis persisted in presenting real evidence that at least some of the components of modern eukaryotic cells, like mitochondria and chloroplasts, were indeed descended from once free-living prokaryotes that long ago began an endosymbiotic union with another cell—whose descendants now include amoeba, ferns, bread mold, and humans. We cover these important ideas again in Chapter 12, because symbiosis is now recognized as one of several important processes involved in evolutionary change.

Evidence is what matters—empirical objective evidence is the currency of science. Faith or dogma should play no part in the scientific enterprise. In practice, this goal is hard to achieve because so much of what science has learned has been stable for so long—and will likely remain so. Even though we take it as fact that the Earth is the third planet from our Sun, we should at least lend an open ear to someone who claimed to have new empirical evidence to the contrary—however strongly we suspect that no such evidence will be forthcoming.

Chapter 2

Natural Selection

Darwinian natural selection was based on a few concepts all obviously true once they had been pointed out. After Darwin had pointed them out, honest biologists agreed they had been extremely stupid not to see them before.

George Gaylord Simpson

The process is remarkably simple. It requires only that individuals of a species vary genetically in their ability to survive and reproduce in their environment. Given this, natural selection—and evolution—are inevitable.

Jerry A. Coyne

Without question, Darwin's most important contribution to our understanding of organic evolution was his concept (discovery) of natural selection. Yes—Alfred Russell Wallace also independently hit on this idea sometime after Darwin, but without the fullness of examples and understanding that Darwin had acquired. When the concept was first presented publically, Thomas Huxley (a great biologist in his own right) is said to have exclaimed something like, "How extremely stupid of me not to have thought of that!" Surely he or other biologists would have come up with this obvious idea had Darwin and Wallace not beaten them to it. There were in fact earlier workers who came exceedingly close to stating the idea, but only in a passing line or two, and without the emphasis and fleshing-out that Darwin was able to provide.

Different writers have since enumerated the fundamental facts behind the process somewhat differently,

Evolution. http://dx.doi.org/10.1016/B978-0-12-800348-0.00002-X
Copyright © 2014 Elsevier Inc. All rights reserved.

but they can be stated as three foundational "facts of nature":

1. Many of the variations in traits within species are heritable.
2. Organisms, and especially sexually reproducing organisms, give rise to variant offspring—offspring that vary from one another in many of their heritable traits.
3. Organisms tend to produce more offspring than their environment can support, so many offspring die (often due to intraspecific competition) before reaching reproductive age.

With these three facts in hand, all that was required was to ask whether the variations present in each new generation have any bearing on which offspring survive to adulthood—and which die young. With just a little knowledge of biology, the obvious answer is a resounding YES— some variations do matter greatly in determining which individuals live or die. This situation results in a "weeding out" of those organisms possessing less favorable variations. Because this is true, natural selection has often been referred to as "survival of the fittest". Some seem to have a problem with that well-known phrase, but if you take fittest to mean what biologists do—best able to survive *and reproduce* in your environment, then survival of the fittest is a relatively accurate description of the process of natural selection.

Natural selection then can be defined as "the nonrandom differential survival and reproduction of genetically variant individuals". To elaborate: some combinations of genetic variations result in more fit organisms (those better able to survive and reproduce) than others, with the nonrandom element saying that the chances of survival and reproduction are due precisely to these resulting genetic variations—not to chance.

So what are the mechanisms of selection that select the unfit? They are in fact many in number and quite varied in their actions. Most probably, any list of selecting factors is going to be incomplete due to our own lack of full understanding. Still, an instructive list would usually include the following factors:

- *Predation*—Except for many photosynthetic organisms, scavengers, and decomposers, most organisms "eat" other living organisms. Lions and hawks eat other animals. Deer and rabbits eat plants. Sea urchins eat algae. Some insects specialize in eating fungi. Many zooplankton eat protists, and many protists eat bacteria. All this can broadly be considered predation, which in most cases means that the organism being eaten will die—one exception being where many plants can recover from losing a few leaves or branches to a herbivore. Even a top-level predator like an African lion can fall prey to a group of hyenas, or a rattlesnake to a roadrunner. Such ever-present biological danger constitutes a significant selective factor for most living organisms. Any variations in organisms that allow them to better escape predation, or better recover from damage done by predators, would on average be less likely to be weeded out by natural selection. Such variations would instead tend to survive and spread in the population.
- *Competition for resources*—Though a few biologists have questioned the relative importance of competition in nature, the logic of ubiquitous competition among organisms is strong. As Darwin and Wallace understood, most organisms tend to produce far more offspring than can actually survive in their environment. Something must be preventing the survival of these "excess" offspring. Add to that fact the additional point that typically all members of the same species require exactly the same environmental resources, and the conclusion that the individuals of many species compete heavily for resources with others of their kind (intraspecific competition) becomes a logical certainty. Competition with members of other species (interspecific competition) can also be an important selective factor. Once again, variations that allow

individuals to better compete for resources will tend to survive and be passed into the next generation. Likewise, the competition inherent in the various forms of sexual selection is an important selective factor that comes into play when males compete with other males for access to one or more females, and in a few cases the reverse. So is the more indirect competition where females survey two or more males before making a choice as to which male she will accept as a mate (mate choice). Competition is obviously a large and complex issue, and it is further fleshed out in Chapter 4.

- *Extreme changes in weather or climate*—As most of us know, even in normal years it is not unusual to have an extremely cold day or two, a month long or longer drought, several days of extremely hot temperatures, floods here and there, damaging hail, and other weather/climate anomalies. Droughts can certainly damage plant and animal communities, as can prolonged extremes of heat or cold. Even sudden "cold snaps" of short duration can kill significant numbers of birds. We can include here the occurrences of rivers that flood or dry up from time to time, since many aquatic species suffer great losses in these extreme situations.

- *Disease and parasites*—So far as we know, every species is plagued by at least one species of parasite—usually more. We can use a broad meaning of the term parasite to include even bacteria and viruses here—any symbiont that negatively impacts its host. In nature, organisms are typically infected with some kind of parasite. I once dissected a young wild-caught opossum in a parasitology lab. I found only one tapeworm in its gut, but also present were at least three types of nematodes in large numbers—certainly over 200 worms in total. Its lungs were also obviously infected with some fungal or protist disease, since it had several yellowish lumpy masses in both lungs. This animal was obviously "diseased" yet it seemed relatively normal on visual inspection. Parasitologists emphasize that this is the norm in nature—for animals to be parasitized at any one time by at least one or more kinds of parasite. So, parasites are a *constant* environmental force selecting "for" organisms that can resist parasitism and survive successfully even under the physiological burden of its parasites.

- *Fire*—In nature in many terrestrial environments, fire can be a regularly occurring phenomenon caused most often by lightning strikes (more recently by humans)—especially in savannahs and mixed scrub grassland environments. Fire directly kills many types of organisms, though some species are adapted to landscapes in which fires regularly occur. Some plants are adapted to survive these regular fires and even gain an advantage when other competing plants are not so adapted.

- *Changes in water chemistry*—For aquatic organisms, water chemistry changes can affect the survival of certain organisms and species. These changes can include dissolved oxygen level, salinity, nutrient levels, silt load, pH, and other factors. Even acid rain can have natural origins, as from the gas expelled during large volcanic eruptions, which in turn can change the chemistry in freshwater systems. Hot weather can lower dissolved oxygen levels. Drought can limit the growth of vegetation in river and stream basins, which can then lower nutrient levels in those streams and rivers.

- *Basic demands of living*—As in the various abilities and structures typically required of the organism, a well-functioning metabolism and physiology, appropriate behavior, effective senses, the required protective structures such as pigments or shells, and various other organism parameters required for normal and effective function within the respective set of environmental conditions. Mutations and/or the inheritance of less-fit alleles can compromise any of these requirements and thus produce individuals that are less fit than the norm. Such individuals would then be expected to more likely succumb to the winnowing forces of natural selection.

There are undoubtedly several more selective factors that contribute to the weeding out of the less fit. That being said, one should keep in mind that natural selection is certainly not the explanation for the death of every single individual. Typically, when ocean fish spawn hundreds to thousands of eggs into the water, most of those eggs may be successfully fertilized. Does then every fertilized fish egg or embryo that gets eaten by ocean predators qualify as a less fit individual? Almost certainly the answer here is no. No matter how successful the genes of a fish embryo might potentially be, there is little an egg or early embryo can "do" to prove its fitness until later in development when it becomes a small fish with behavior and traits that might lead to its success. The earliest developmental stages of fish, especially those before hatching, have little opportunity to prove their fitness during a time when many mouths are hungry for those nutritious morsels. Only genetic defects causing malfunctions in early development could be weeded out by selection at this time.

This situation also extends to plants. A pine tree growing near a lake automatically has a disadvantage (no matter how fit its genes) since many of its released seeds will land in the water and likely die—rather than in soil where their continued existence might be possible. In these and many other diverse situations, unimaginably vast numbers of young organisms on this planet die due to what we might call bad luck rather than any lack of genetic fitness.

Over the long history of life on Earth, many millions of individuals, and even species, have died out as the result of chance cataclysms like the impact of asteroids, large-scale volcanic eruptions, unusual and drastic climatic swings, etc. The dinosaurs that were still surviving before the last major extinction event some 65,000,000 years ago were probably mostly well-adapted as species and individuals to their environment, but cataclysms of this magnitude (at least in part due to a large asteroid impact) do not take note of or respect adaptations selected for successful living in "normal times". Again, though such events seem rare, the magnitude of their exterminating effects serves as another reminder that natural selection is certainly not the only cause of death on this planet.

One more restricted example of death by chance rather than selection occurs in a few species of birds that exhibit the phenomenon of siblicide. In the black eagle of Africa, females typically lay two eggs a few days apart, which likewise then hatch a few days apart. This results in the older chick having a 3–4 day head start in feeding and growing over the second hatched chick. In almost all cases, within the first hour of life of the second chick, the older chick initiates an ongoing series of pecks and aggressive behavior directed toward its younger sib—usually resulting in the death of the younger chick within a day or two (Mock et al., 2010). This general phenomenon is common with some variations in ospreys, blue-footed boobies, and a few other bird species.

I mention this siblicide phenomenon again in Chapter 4 as an interesting example of competition, but it is not necessarily survival of the fittest in terms of genetic fitness—and therefore not natural selection. Unless the first chick to hatch had major genetic defects, it would be the larger and stronger of the two chicks hatched and would be able to dominate and dispatch its younger sib. In other words, it is entirely possible that the second hatched chick might be genetically superior to the first, but due to the chance order of laying and hatching, the older chick is easily able to dominate and kill the younger. To use an analogy, no matter how fast a runner might be, if a slower competitor has a 60 m head start in a 100 m race, the slower runner would likely win. There are undoubtedly a great many other specific circumstances in the millions of species out there where chance, timing, or bad luck becomes an important factor contributing to the death of individuals.

Still, most young organisms that do survive beyond these early stages enter a time in their lives when numerous selecting factors like those listed earlier will enter the picture. To emphasize again, natural selection is a weeding-out process in which individuals better suited (because of their variations) to survive and reproduce simply outlive and out-reproduce those with less successful or "fit" variations. There is no selecting power or agent that actively selects and favors certain variations, only a tough environment with many challenges that will not be successfully navigated by all. As the great evolutionary biologist Ernst Mayr once put it: "Actually the *selected* individuals are simply those who remain alive after the less well-adapted or less fortunate individuals have been removed from the population" (Mayr, 1997). Stephen J. Gould reiterated this point when he wrote: "Selection carves adaptation by eliminating masses of the less fit—imposing hecatombs of death as preconditions for limited increments of change" (Gould, 1994).

This point is important to a firm understanding of evolution. Many mistakenly believe that if "something" is being selecting, there must be some desired outcome of the selection process. This misconception often goes hand-in-hand with another big misconception that evolution is a progressive process that innately leads to more complex and perfect types of organisms. Some even believe that intelligence and the humanoid form are goals of the evolutionary process. These ideas are all at odds with our modern understanding of evolution. On the intelligence issue, Bertrand Russell once insightfully wrote: "If it is the purpose of the Cosmos to evolve mind, we must regard it as rather incompetent in having produced so little in such a long time." Again, the only goal of the evolutionary process—and it is really more a result than a goal—is the production of organisms capable of surviving in their environment and successfully passing copies of their genes into the next generation. Bacteria are still obviously very capable on this score, as are amoeba, ferns, and millions of other groups that have relatively low complexity and/or intelligence. It is also doubtful that trilobites in their day were in any way less successful creatures than shrimp are today. The last trilobites became extinct during the great Permian extinction event, which wiped out around 90% of all the species on the planet. Again, overall general fitness almost certainly has little to do with surviving or being wiped out in the face of such a catastrophic event. There are now various ideas as to what factors caused this greatest of all extinction events, but those are the topics of other books.

Natural selection is traditionally divided into three general categories titled stabilizing selection, directional selection, and disruptive selection. The first of these—stabilizing selection, is almost certainly the most common form of selection occurring in nature. Stabilizing selection occurs whenever a successful trait occurs in the great majority of the individuals comprising a population, and there is currently no better or more successful variation of that trait. If this is the case, then most of the variations from this successful norm that occur (as they likely will in some of the variant offspring) will be selected against, which will in effect maintain the status quo with most individuals possessing that successful trait.

Most of the thousands of traits in each species are under stabilizing selection most of the time. If it were otherwise, the population would be generally unfit and would likely be on a fast track to extinction. However, there is almost always a small subset of traits that are under the other form of selection—directional selection. As well-adapted as humans may be, we still have a few traits that are under at least weak directional selection, with the most obvious being our several useless vestigial structures. No one today needs their wisdom teeth—as we did when early humans had longer and more "ape-like" jaws. Wisdom teeth typically now come in (erupt) much later than our other teeth (and in some individuals they never do so at all). They also often cause pain and crowding of the other teeth, pressure on facial nerves, and possible infection—plus being a complete waste of nutrients and calories to construct.

Before dentists came along, humans experiencing wisdom teeth problems really were less successful than those whose wisdom teeth never erupted, or came in much later in life, or never formed at all. Some likely died of resulting oral infections, or were distracted by the pain and so were easier prey for predators, or could not eat as effectively. As a result, those with active wisdom teeth were at a disadvantage and were thus less able to survive and pass their genes on. It may seem hard to believe that something like wisdom teeth could be a factor in selection, but over hundreds of generations in a natural setting, they certainly would have been a negative and therefore would have been selected against.

In some cases, new traits that generate completely new structures are favored. The mammal line leading to whales and dolphins is now known to have started from a relatively small terrestrial artiodactylid. DNA analysis recently confirmed modern hippos as the closest living relatives of whales. Most modern artiodactylids like deer, camels, and goats have small but somewhat typical mammalian tails, whereas dolphins and whales have evolved horizontal lateral flukes at the end of their tails, a trait approached independently by only one other mammal group, the manatees. As the early ancestors of whales moved further and further toward an aquatic existence, these structures somehow evolved, and in this case the nature of the genetic variations involved in this transformation is still unknown, but the fact remains that these new structures did evolve (probably gradually) from tails that previously were typical mammalian tails. As variations created any flattening and widening of the tail, swimming ability due to up and down undulations of the tail was enhanced, and those variations were maintained while the older normal tail variations lost out.

The third category of natural selection is known as disruptive selection. In disruptive selection some character (or set of characters) in a population spans a range of variation, and the fittest phenotypes within that range of variation happen to fall near each of the two extremes of the variation range, with the intermediate forms being the ones selected against. Generally this would be like the medium-sized individuals being selected against while the

larger and smaller are both sufficiently fit to escape selective removal. This would lead to a population characterized
by larger and smaller individuals, but few or none in the
medium range. Such a species would be termed polymor-
phic in that more than one phenotypic optimum is present,
and more than one norm for that phenotype is common.
What are some examples of this rare form of selection?
Known examples of this type were once few in number.
Sexual dimorphism is sometimes used as an example of
this phenomenon, where one sex looks and behaves in one
way—and is often larger, while the other sex looks and
behaves differently. This would not be true of asexual or
monoecious species, but is very obvious in peacocks, ele-
phant seals, and African lions.

Actually, the best examples of disruptive selection now
seem to be in the several apparent cases of sympatric speciation, where two subgroups of a local population start to
diverge in some characteristic(s), eventually to the point
that the two groups look or behave differently enough to
lead to speciation of the two types (see Chapter 13 for specific examples).

Yet another subcategory of natural selection is termed
balancing selection. In balancing selection, two or more
variations of the same gene or trait are favored and maintained in the population. The classic example of balancing
selection is that of the maintenance of the sickle cell hemoglobin allele in parts of Africa where malaria is prevalent.
Sickle cell of course is a harmful genetic disease. People
who have inherited two sickle cell alleles from both parents have the disease, and without lifetime medical intervention they would likely die in youth. Even heterozygous
individuals with one normal and one sickle cell allele are
disadvantaged and lack the stamina and vigor of someone
with two normal hemoglobin alleles. So why is the sickle
cell allele favored? It is favored or maintained in these areas
because those heterozygous individuals are far more resistant to malaria than people with two normal hemoglobin
alleles. That is, malaria is more life threatening than being

heterozygous for the sickle cell gene, so malaria tends to kill more of those lacking a sickle cell allele. In those areas where malaria is common, those with two sickle cell alleles are at great risk of death from their sickle cell disease, while those with two normal hemoglobin alleles are at great risk from malaria. Those, however, with one normal and one sickle cell allele are spared the harshest selection from either of these two factors—therefore the selection is balanced in maintaining both alleles in those populations.

Another major contribution from Darwin was his recognition and elaboration of the concept of sexual selection. Wallace never had much to say on this topic and did not accept its importance in evolution, while Darwin wrote extensively on sexual selection, especially in his later book *The Descent of Man and Selection in Relation to Sex* (1871). Surprisingly, the field of biology did not really immerse itself in studying this important aspect of evolution until well into the 1900s. Some workers have mistakenly treated sexual selection as a phenomenon separate and somehow opposed to natural selection. Natural selection was viewed as concerned with promoting survival, while sexual selection was concerned with promoting reproduction.

Such thinking is wrongheaded, since survival without reproduction counts for nothing in evolutionary terms. Mules as individuals have good survival value and physiological stamina, but in terms of evolution, they are big zeros since they are sterile hybrids between donkeys and horses and pass no genes into the next generation. In nature, organisms ultimately survive only for the purpose of reproducing or gaining genetic fitness, and therefore survival and reproduction cannot be divorced from one another or considered as separate evolutionary goals.

Sexual selection is a huge topic and is itself the topic of several books and many hundreds of articles. Hopefully it will be enough here to emphasize again that sexual selection is a category of natural selection more immediately related to reproductive success. To put it briefly and simply, some of those genetic variations within species will affect the ability of some individuals to locate a mate, to choose a good mate, to successfully court or compete for a mate, to hold and protect a mate, etc. This is the realm of sexual selection. Only in asexually producing organisms like amoeba would this aspect of natural selection be absent. Specific examples of sexual selection can fall under stabilizing, directional, or disruptive selection, or any combination of these. As mentioned earlier, sexual dimorphism within a species can be said to have resulted from disruptive sexual selection.

One further topic of importance in a discussion of natural selection is the question of what exactly is the "unit of selection" in natural selection. Is it the species, or an even higher taxon like the genus or family? Most biologists do not believe so. Is it a group of organisms such as a population or smaller group? Most biologists again believe this is not generally the case, with one possible exception being the colonies of social insects such as ants and bees, in which only one or a few individuals do the reproducing, while the other colony members are sterile. In some sense these "superorganisms" do form a reproductive unit analogous to an individual organism of most nonsocial species, and may be said to compete against other intraspecific colonies (Holldobler and Wilson, 2009).

Most of the discussion so far in this chapter has appeared to indicate that individual organisms are the units of selection, especially since they are the units that die young, or survive to reproduce, or have varying degrees of success in achieving genetic fitness. As obvious and logical as this appears, many biologist now view the genes themselves as the units of selection, and this was the major theme in Dawkins' revolutionary book entitled *The Selfish Gene* (Dawkins, 1989). An easy way to illustrate this point is to contrast natural selection with artificial selection in which people are selecting for certain traits or combinations of traits in plants or animals. Such breeders are not selecting for individuals (you can correctly say selecting "for" in artificial selection since the breeder is doing just that) since individuals live only once. They instead are selecting for particular traits and combinations of traits they hope to establish in future generations of organisms. Remember that traits are coded by genes and combinations of genes, so it is really those genes that are being selected in cases of artificial selection. In natural selection, much the same thing is going on, except that the selection is negative against genes and gene combinations that produce less fit traits in individuals, while the more fit traits tend to survive (in fit individuals) and be perpetuated in the next generation more successfully. Many additional examples that support this viewpoint are to be found in several of the books by Dawkins.

I end this chapter by pointing out that several scientists and writers have stated that Darwin's discovery of natural selection (the prime mover of evolution) is arguably the greatest scientific insight or discovery of all time. I would have to agree with this assessment because of the all-encompassing nature of this insight and the vast number of questions that are answered in whole or in part by this line of thinking. It is simply one of the few key ideas that has allowed science to make sense of much of nature. In short, it explains a lot.

Chapter 3

Adaptation

Every species is pretty well adapted, which means that selection has already brought it into sync with its environment.

Jerry Coyne

Natural selection tends to produce organisms that are "adapted" to their local environments. Each organism is in large part a collection of many distinct adaptations that allow it to function well (survive and gain genetic fitness) in its environment. An organism that can do this effectively is said to be a genetically fit or well-adapted individual. So what exactly is an adaptation? An adaptation is a heritable trait that has been "selected for" in ancestral populations because of its positive contribution to genetic fitness. Again, a better wording would be any trait that is inherited genetically that has been retained (not weeded out by natural selection) in a species because of its past positive effects on fitness. The adaptations that an organism inherits are always those that worked well in previous generations. Evolution cannot foresee and plan adaptations in the future, so if environmental conditions change suddenly, the new generation will most likely not be so well-adapted to those new conditions, and natural selection then will become more strongly directional to move the appropriate traits (if possible with the available genetic variation) into better accord with the new environmental conditions.

Adaptations span a range of categories that include biochemical, physiological, morphological, and, in some groups, behavioral. The great majority of adaptations

Evolution. http://dx.doi.org/10.1016/B978-0-12-800348-0.00003-1
Copyright © 2014 Elsevier Inc. All rights reserved.

originate from one of three sources: modifications of existing characteristics (including exaptations), the addition of new characteristics, or the loss of former characteristics.

The modification of an existing trait can take two forms. A trait can simply be refined to better serve its current role. For example, the fur of a mammal may serve to keep the animal warm in cold weather. If the climate changes slowly to a colder norm, selection should favor longer or denser fur to better insulate the animal—the trait has the same function, it is just modified to better serve that function. There are many examples of this kind of trait modification, such as wing traits for better or different types of flight in birds, increased attractiveness to insects by the flowers of insect-pollinated plants, and increased efficiency of camouflage in many animals.

Alternatively, evolution can adapt an already present characteristic to an entirely new function that adds fitness value in a new way, a change known as an *exaptation*. Lobe-fins evolving into limbs during the fish-to-amphibian transition was an exaptation—the modification of fins into limbs. True, this modification still enabled locomotion, but of a very different kind. Forelimbs in tetrapods were then later exapted into wings in birds, in bats, and also in the extinct flying reptiles, the pterosaurs—again a distinctly new use of the forelimbs. In flowering plants, some photosynthetic leaves exapted into the nonphotosynthetic petals of the flowers. In the venus flytrap, some leaves exapted into insect traps and functional "stomachs" for the digestion of insect prey. Some of the snout hairs of the early mammals were exapted into sensory whiskers. Cartilage supporting gill openings in some early fish were exapted into the first vertebrate jaws. The "venom genes" of modern venomous snakes are typically a mix of genes formerly forming components of saliva and genes that were active elsewhere in the body—all now exapted (often after duplications) into venom genes (Zimmer and Emlen, 2013). These comprise but a small sampling of the vast number of examples of exaptations that have occurred in the evolution of life on Earth.

Most adaptations involving the formation of new characteristics occur through exaptations of existing genes and structures, yet in some cases adaptations can be something entirely new at the genetic and trait level. Perhaps the most clear-cut examples of this sort of adaptation occur in prokaryotes. It is now known that through the process of horizontal gene transfer (explained further in Chapter 7) one line of bacteria or archaea can receive entirely new genes from another line of the same or a different species in the form of a plasmid. New traits associated with these newly acquired genes can allow an increased ability to parasitize other organisms—so-called virulence factors. Some transferred plasmids can convey the ability to metabolize new nutrient sources or survive under different environmental conditions. There are at least a few examples of what seems to be horizontal gene transfer from prokaryotes to eukaryotes, which have instantly resulted in new genes and traits in the eukaryote (more detail is given in Chapter 7).

We tend to think of several of the adaptations in eukaryotes as completely new inventions, such as hair in mammals, feathers in birds, viviparous birth in "higher" mammals, and tail flukes in whales. Even these can involve some level of exaptation of already existing genes and traits, although in many cases this is hard to investigate or confirm since the traits in question appeared millions of years ago, leaving no clear or discernible record of their origins.

Although often neglected in discussions of evolution, adaptation is often accomplished through the loss of formerly useful traits and features. There are an estimated 2,500 species of fleas in existence, and they all descend from an ancestor that lost the typical insect wings when adapting to an ectoparasitic lifestyle spent largely in the thick fur of mammals, where wings would be of no use and possibly even hinder their movements. Similarly, some 1,100 species of tapeworms have evolved from an ancestral flatworm group that lost the digestive tract since it was unnecessary for worms living in the gut of vertebrates, where abundant nutrients were already present in predigested form and only needed to be absorbed through the worm's outer body surfaces.

Apes (including humans) have lost their tails except for a few vestigial caudal vertebrae. We humans have "lost" (actually just significantly reduced the size of) most of our body hair. Several completely parasitic plant species such as the famed *Raffelzia arnoldii* (famous for producing one of the largest of all flowers) have lost their ability to form chlorophyll and carry out photosynthesis. *R. arnoldii* is now heterotrophic, getting all its nourishment from the host plant (mostly tropical vines). The kiwi bird of New Zealand has all but lost its now tiny wings. One cannot even see the small vestigial wings of a kiwi without holding one down and searching among the hair-like feathers of the shoulder region for the tiny remnant wing. Other island-dwelling birds like the kiwi have independently lost functional wings because there were no predators on the islands to attack them; although you might think that flight would be useful for many purposes, it has been independently lost over and over again by several island-dwelling bird species in areas where the threat of terrestrial predation was absent, showing that the main function of flight is escape from predators on the ground. The now extinct dodo bird (a relative of pigeons) of the island of Mauritius is another example of this, as is the kakapo (a kind of parrot) found also on New Zealand. These latter two had/have obvious but reduced wings, and the birds were/are incapable of flight.

Even organelles can be lost. It was once believed that some protists such as *Giardia lamblia* (an intestinal parasite that can cause diarrhea in humans) lacked mitochondria primitively; that is, it was descended from an early line of protists that had never acquired mitochondria through an ancestral endosymbiotic event (see Chapter 12). Today we have evidence that *Giardia* species and their kin are in fact descended from a line of protists that once had mitochondria but apparently lost them far back in evolutionary time when adapting to anaerobic environments (the intestinal lumen in the case of *Giardia*). The evidence that *Giardia's* line once had mitochondria comes from its nuclear genome, which still contains DNA of mitochondrial origin—similar to the mitochondrial DNA found in other eukaryotes that still have mitochondria—but in which many of the original mitochondrial genes have relocated into the nucleus from the mitochondria (Lane, 2005).

Another recently discovered example of organelle loss involves the loss of functional chloroplasts in many members of a clade that includes the apicomplexans, the ciliates, and a few other related groups (Huang and Kissinger, 2006). Some of the protists in this supergroup (the alveolata) retain a remnant of the ancestral chloroplast but with no or reduced functionality (none retain the ancestral photosynthetic function).

The apparent loss of a trait can result from the actual loss, or severe mutation, of the gene(s) or promoter gene(s) associated with that trait. In most cases of trait loss, the coding gene(s) still exists in the genome in a mutated and inactive form, where it may continue to exist for thousands of generations. A completely inactivated gene is known as a *pseudogene*. Only an accidental deletion of the now-functionless pseudogene could remove it completely from the genome.

Adaptations are recognized as such because they are known to increase in some way the fitness of the respective organism. *Fitness* is a term with a common usage denoting the physiological fitness of an individual's body—muscularity, power, and endurance. In biology, however, the term *fitness* is shorthand for *genetic fitness*, and this has a very different meaning from the common body-fitness usage. An organism can be considered fit if it is *likely to be able* to survive in its environment to reproductive age and then successfully reproduce a number of equally able offspring. A more technical and correct way to put this is that if an organism is fit, it is well equipped to be able to pass on and promote copies of its genes into the next generation.

To briefly contrast these two usages, mules have long served as a useful illustration. In terms of bodily fitness, mules are typically muscular and have good health and stamina. That is why they were once commonly used as work animals to pull plows and carry heavy loads. However, as the result of a donkey/horse cross, mules are not a true species of animal, and to add to their strangeness, they are sterile (unable to produce offspring). So, mules do have great fitness in the sense of bodily fitness, but they have no fitness whatsoever in the biological sense since they cannot pass on their genes to another generation.

Evolution can aim at only genetic or biological fitness; it is obvious that if genes are not passed on to offspring, those genes cease to exist. The world is by default populated with organisms whose genes have been passed down through generations because those genes successfully promoted genetic fitness in their ancestors. There are many organisms whose lifestyle has demanded that bodily fitness be present as well to help with the "survival until reproduction" component of life. Many antelope have amazing stamina and muscularity which allows them to escape predators—and so they can survive into adulthood and reproduce. On the other hand, many animals exhibit little in the way of bodily fitness because it is not required for their lifestyle and survival.

Almost no one familiar with tapeworms would claim that they were fit in the sense of power and stamina, but tapeworms are in fact very fit animals in terms of biological fitness; they have been around for many millions of years and continue to persist as parasites in a vast array of vertebrate hosts. Also, *fitness* is a term that can be applied to any successful living organism including plants, mushrooms, and bacteria—all organisms to which no one would relate the common usage referring to bodily fitness.

Biologists more often use the term *fitness* not to contrast between different species, which are generally all very fit, but to compare the relative fitness of different individuals within the same species or population. Any individual carrying a lethal gene or allele and who is still alive would be judged to have lower fitness than an individual not carrying that particular lethal gene (all other things being equal). Likewise, an individual with little resistance to disease would be less fit than one of the same species who had a higher level of disease resistance. An individual lacking attractiveness to the opposite sex could be judged to have low potential fitness, which relates back to sexual selection (Chapter 2).

So, adaptations are commonplace among all organisms. They are the products of natural selection and they occur in many different forms. As environments change, some traits that were good adaptations in the past may become detriments to survival and reproduction and so will be selected against. In short, some adaptations will be short-lived in terms of geologic time, while others will continue their successful effects far into the future. This chapter has certainly given the impression that most genes and traits are adaptive and that organisms are essentially just big collections of adaptations. In fact, this is not the case. Most of life is riddled with imperfections—the topic of Chapter 16. They also often have genomes that are considerably less than fine-tuned and efficient operating systems, as we will see in Chapter 5. Evolution, it turns out, is a somewhat sloppy process, resulting as much from chance as it does from the accumulation of useful adaptations.

Chapter 4

Competition

As the individuals of the same species come in all respects into the closest competition with each other, the struggle will generally be most severe between them.

Charles Darwin

Natural selection invariably turns variation into competition.

John Tyler Bonner

Competition in nature takes many forms and occurs at many levels. Many people have the mistaken idea that competition typically involves direct confrontation as in the case of male lions fighting for control of a pride, or songbirds battling it out for control of a territory. Competition can often be less than obvious in nature, often occurring in indirect ways and over protracted time spans that do not result in any obvious "confrontations." Biologists generally agree that competition for light is the prime reason why any plants grow more than a few inches in height—to grow above competing plants and thus intercept more sunlight. Below ground, plants also compete for soil nutrients and water. If a plant disperses its seeds by the wind, and a new variation in seed design allows for longer dispersal distances, the plants producing such seeds will—all other things being equal—have a competitive advantage in getting their seeds dispersed far and wide, and will therefore have a better chance that some of those seeds will reach a suitable habitat for germination and growth. Plants compete in a great many ways that are devoid of the bloodshed or aggression

Evolution. http://dx.doi.org/10.1016/B978-0-12-800348-0.00004-3
Copyright © 2014 Elsevier Inc. All rights reserved.

present in some animal examples. There are however a few examples where plants can be downright nasty in their competition. Some plants are known to secrete chemicals from their roots and/or leaves that directly kill any closely neighboring plants of other species. The black walnut tree *Juglans nigra* is a well-known example of this type of chemical warfare, secreting a chemical that will kill most other plants within close proximity (Scott, 2008). This certainly does compare with the more violent of competitive encounters in the animal kingdom, but of course plants don't even know they are competing—even though they are.

Some birds compete for limited nesting sites simply by finding them first and occupying them before others. The old phrase "survival of the fittest" implies competition among competing entities, whether they be groups, individuals, or genes/alleles. One of the few facts of nature that led Darwin and Wallace to their concept of natural selection was that in all known species, more offspring are produced than can possibly survive in the available environment, with the average situation being that for every two parents, two offspring *of that species* will survive to adulthood to replace them. The two replacement offspring need not be from the two parents they are replacing. Some parents will leave one or no offspring, while others will leave more than two. Of course those leaving two or more would have outcompeted those leaving one or none. Here then is another example of competition not necessarily involving direct aggression or confrontation.

Since individuals vary in their genotypes and resulting phenotypic traits, some will be more able to survive and produce the next generation than others. Since this is not a random outcome (who lives and who dies young—who reproduces successfully and who does not), the obvious reason why some fail while others thrive is that the "fittest" outcompete the less fit in some way. Perhaps they are faster, perhaps their senses are more acute, perhaps they are better camouflaged, perhaps they mature earlier, perhaps they have a more efficient metabolic pathway, or perhaps they are better parents. These and thousands of other differences can result, depending on the species, in a competitive edge in terms of survival and attainment of genetic fitness.

Two individuals of the same species typically need *exactly* the same resources from the environment. This is generally less true for individuals of different species. Two tigers need and compete for the same exact resources in their environment—a tiger and an earthworm compete for very little. This means that intraspecific (within the same species) competition should typically be a big component of the "struggle for existence." A dramatic example of this is the phenomena of siblicide, in which one newborn kills its own sib or siblings. Some birds engage in this behavior, and in some the killing is a regular event occurring each breeding season. The black eagle of northeastern Africa is known to be an obligate siblicidal species. Typically two eggs are laid a few days apart, and the first chick to hatch almost always kills the second hatchling. In this species, the reason why two eggs are produced is believed to be as insurance in the case that one of the two eggs might not hatch (a not uncommon occurrence in birds). Also, in this species the parents would typically not provide enough food to fledge two chicks, so natural selection has favored innate siblicidal behavior in the chicks (Mock and Drummond, 2010). A greater number of bird species engage in facultative siblicide, where killing may occur only in "bad years" when food is limiting. In good years when the parents are able to deliver more food to the chicks—therefore reducing the competition—all the chicks may survive to adulthood and thereby increase the fitness of the parents.

Most known antibiotics are produced by microbes and fungi as a tool to reduce competition. The antibiotic is not lethal to the species of microbe producing it, only to other microbes which might try to crowd in and utilize the same food resources as the antibiotic-producing microbe.

Fortunately, humans learned to harness this potential health tool that existed in nature due to microbial competition.

One of the five to six main reasons given for the huge success of insects as a group is that most insect species have evolved complete metamorphosis, which almost always results in a reduction or even total lack of competition between the young insects and the adults (Daly et al., 1998). In mosquitoes for example, the aquatic young live in the water and feed mainly on algae, while the winged adults live in the terrestrial environment and feed on blood or nectar. Caterpillars feed mainly on leaves, while the adult moth or butterfly feeds on nectar. The most diverse insect orders are in fact those that have evolved this type of development where competition for food is essentially abolished between the young and the old. This one point alone is strong circumstantial evidence that competition is a large and overriding force in nature, when one considers that at least one out of every five known animal species is a beetle, and beetles are one of those insect orders that have complete metamorphosis.

The list of obvious examples of competition in nature is a very long one. In many animals there is competition for mates, which typically drives males to do battle to win females, establish and defend territories, build impressive bowers in the case of bowerbirds, engage in costly courtship displays, be burdened down with costly ornamentations like the tail of male peacocks, or even guard the female after mating to prevent another male from displacing the sperm of the previous male. Corals are in continual competition for space on a coral reef, both with other corals and with sponges and several types of algae which require attachment to the substrate. When an overcrowded honeybee colony splits into two, the departing queen leaves behind two or more larvae that will develop into new queens, but since a hive can only have one queen, the first of the new queens to hatch out will immediately locate the other one or two queen cells and kill her rivals—pretty obvious and harsh competition in this case.

Even historically there is evidence that competition was a powerful force in shaping nature. When a variety of North American placental mammals first gained easy access into South America across the newly exposed Panama land bridge around 2,700,000 years ago, they apparently out-competed and drove to extinction a significant number of the endemic marsupial species there. Farther back, mammals themselves appear to have been held in check in terms of diversification and size increase during the reign of the dinosaurs—only diversifying significantly and increasing in size after the demise of the dinosaurs 65,000,000 years ago. Whales evolved only after several predatory niches were vacated at about the same time when the many kinds of ichthyosaurs, plesiosaurs, mosasaurs, and other large predatory marine reptiles became extinct. In these last two cases, when competition was reduced or eliminated, the door of opportunity opened for new adaption and the speciation of surviving groups. Another historical example of competition closer to home seems to be the eventual triumph of *Homo sapiens* over *Homo neanderthalensis* in Europe, where competition is the answer of consensus for the extinction of Neanderthals among those studying human evolution (Zimmer, 2005).

A few biologists have downplayed the extent and importance of competition in nature. Most of those that have done so argue that cooperation is as common, or even more common, than competition, and that nature actually "seeks" cooperation and harmony. Even among biologists, people often see the same physical reality and interpret it differently depending on their experience and understanding of nature. It is true that there are countless examples of cooperation in nature, both within and among species. There are the many known examples of mutualisms such as insects pollinating plants, protists aiding in the digestion of wood in termite guts, micorhizal fungi aiding in nutrient uptake by plants, and zooxanthellae living within hermatypic coral polyps and sharing an interchange of nutrients. Though examples of intraspecific cooperation and mutualistic relationships between species are abundant, the logic of ubiquitous competition among organisms remains strong. To repeat again Darwin's fundamental point—most organisms tend to produce far more offspring than can actually survive in their environment. Something must be preventing the survival of these "additional" offspring. Add to that fact the singular point that all members of the same species require exactly the same environmental resources and the conclusion is almost certain that individuals typically compete heavily for resources with others of their species (intraspecific competition).

Often, there is additional interspecific completion (competition with members of other species) for most of these resources as well (leaving out that for mates of course). Lions and hyenas in Africa most definitely compete for space and food, with competitive encounters over prey kills occurring regularly, and outright killing of hyenas by lions—and of lions by hyenas occurring on occasion. Competition is a driving factor in the process of ecological succession (especially among the plants), and it is the most obvious factor in the success of many introduced species which in one way or another are able to out-compete some of the native or endemic species in that environment. In short, competition appears logically to be ubiquitous and multifaceted throughout nature.

Still, some stress cooperation over competition; likely it is a case of subjective interpretation. Take the case of the mutualism between zooxanthallae (photosynthetic dinoflagellates) and their host hermatypic corals (any hard coral that hosts zooxanthallae) as an example. Have these two *unconscious* types of organisms "sought out" this mutualism? Was there some tendency or force encouraging cooperation for cooperation's sake? Almost certainly not, at least there is no objective evidence that this could be the case. Cooperation usually comes about *because of* competition. When this mutualism was in its formative stages, those corals that started to house zooxanthallae derived benefits that allowed them to better compete with corals lacking this cooperative association, and initially this would have involved other corals of the same species as well as corals of other species. If corals housing zooxanthallae grow faster and gain additional nutrients from their cooperation with zooxanthallae, they will tend to outcompete those corals lacking these benefits. Over many generations, the entire population will become composed of corals having these mutualistic properties because those unable to utilize this *competitive edge* will have lost out in the harsh completion for survival. Richard Dawkins has argued (Dawkins, 1998) that cooperating "members" of such mutualisms are in fact "selfish cooperators" because cooperation is most often established as a way to "beat the competition" and allow both members of the mutualism to thrive—again, it never was cooperation for cooperation's sake, or the result of some harmony-seeking drive in nature (which some have actually argued for, even though this is not a scientifically-based premise). Most biologists fully understand the harshness of nature and the many competitive threats to survival and genetic fitness.

The most pressing and ever-present evolutionary pressures affecting species and their evolution are not climate change, continental drift, ocean level changes, or other such slow and large-scale abiotic changes. They are rather the day to day presence of biotic competitors and other potential threats such as disease, parasites, and predators. To emphasize this important point and to rebut again those who believe in a more cooperative and harmonious Nature, allow me to end this chapter by sharing these two telling quotes:

- "One of the most appalling realizations with which every student of nature is brought face to face is the universal and unceasing struggle for existence which goes on during the life of every living organism, from the time of its conception until death. We like to think of nature's beauties; to admire her outward appearance of peacefulness; to set her up as an example for human emulation. Yet

under her seeming calm there is going on everywhere—
in every pool, in every meadow, in every forest—murder,
pillage, and suffering" (Chandler, 1940).

- "A forest or a coral reef under a blue sky and bright sun-
shine gives an impression of tranquility and harmony, but
the impression disappears when details are examined.
Look at one of those trees in the forest. Almost certainly
you will find that it is afflicted with pests and diseases
and is under frequent attack by browsers such as deer or
howler monkeys. The same can be said of those monkeys. Causal examination of their skins may show the
ravages of fleas or ticks or fungi. They live in constant
danger from attacks by jaguars and other predators. The
story of the forest or coral reef is a tale of relentless arms
races, misery, and slaughter" (Williams, 1996).

Chapter 5

Genetics Basics & Mutations

One can never fully understand the process of evolution unless one has an understanding of the basic facts of inheritance, which explain variation.

(Ernst Mayr)

The fact that the replication of DNA is not without error is the root of evolution.

(Peter Atkins)

As was mentioned in Chapter 1, the current definition of evolution focuses specifically on genetic or genomic change over time in populations of organisms. Another wording often encountered refers to changes in the *gene pool* of a population or species—another bit of jargon that refers to the total number and frequency of genes and alleles in all members of a population or species at any one moment in time. It is mainly the heritable genetic material of *some* individuals that is copied and passed on from one generation to another, so it is the total "pool" of these heritable entities that evolves—and I am referring here to the entire genome as was discussed in Chapter 1. Darwin and Wallace knew nothing of the molecular basis of genetics, but they did know (Darwin especially) the power of heritable variation in originating change in organisms.

Today we understand vastly more about inheritance and genetics than Mendel or anyone else back in Darwin's day—and we are still on a learning curve in these topics. Because of the relatively recent discovery of several varied processes that can alter and affect DNA (and thus

Evolution. http://dx.doi.org/10.1016/B978-0-12-800348-0.00005-5
Copyright © 2014 Elsevier Inc. All rights reserved.

genomes and inheritance over time), a full comprehension of genomic evolution has become a daunting goal. We will hopefully cover here and in other chapters enough of these topics to illustrate most of the diverse mechanisms that can affect genetic/genomic changes over time.

Though the simple monohybrid and dihybrid crosses covered in basic biology courses are great oversimplifications of the actual depth of genetic variability and complexity, they do nevertheless illustrate a few central facts concerning inheritance. Refer to the dihybrid cross in Figure 5.1 and note the following:

1. Gene alleles are inherited from the two parents (in sexually reproducing species).
2. Some of the possible resulting offspring have genotypes and phenotypes that differ from either of the two parents.

	WT	Wt	wT	wt
WT	**WW TT** red tall	**WW Tt** red tall	**Ww TT** red tall	**Ww Tt** red tall
Wt	**WW Tt** red tall	**WW tt** red short	**Ww Tt** red tall	**Ww tt** red short
wT	**Ww TT** red tall	**Ww Tt** red tall	**ww TT** white tall	**ww Tt** white tall
wt	**Ww Tt** red tall	**Ww tt** red short	**ww Tt** white tall	**ww tt** white short

FIGURE 5.1 A simple dihybrid cross involving two genes with two alleles each—in which: W=red flowers, w=white flowers, W is completely dominant over w; T=tall plant, t=short plant, T is completely dominant over t. Parent one is WwTt with red flowers and tall growth. Parent two is WwTt with red flowers and tall growth.

3. Several genotypes and phenotypes are possible outcomes for the offspring of just two parents with their two respective genotypes and phenotypes.

These facts, illustrated by even a simple dihybrid cross, include two of the three essential facts on which Darwin and Wallace founded their concept of natural selection—those being: (1) that some traits are heritable, and (2) that sexual reproduction results in offspring that typically vary from each other—and from their parents.

When you step up to a cross examining the inheritance of three genes, each with two alleles, the amount of possible variation in offspring increases, and an even smaller percentage of possible offspring would share a genotype or phenotype with either parent. Add co-dominance for some allele combinations or multiple alleles for some genes and the degree of possible variation in offspring increases again. Then consider that most organisms are coded for by thousands of genes—many of which have two or more alleles, and the potential variation in offspring moves to a level that is simply beyond comprehension.

Even if you can somewhat comprehend the amount of potential variation resulting from only allele shuffling and recombination in gametes and offspring, this is still only the tip of the variation iceberg, since in cases of multiple alleles (three or more alleles per gene) there may exist complex assorted dominance relationships—with some alleles of a genes showing complete dominance with particular other alleles and co-dominance with yet other alleles of the same gene. An example of co-dominance would be if the genotype Ww (from the cross in Figure 5.1) were to result in pink flowers rather than either red (as WW would have produced) or white (as ww would have produced). Additionally, many traits are known to be polygenic because they are directly affected by two or more genes, each of which may have two or more alleles. Also, there are many possible intergene affects (pleiotrophy) where a particular phenotype can be affected by genes other than the "primary gene" associated with that trait. Though we need not detour here for what would amount to pages of explanation and examples on these several generators of variation, suffice it to say that variation is typically vast and complicated in sexually reproducing organisms. If some variations in such traits decrease the potential fitness of the individuals who possess those variations, those individuals will on average be selected out of the population at a greater rate than individuals carrying more successful forms of those traits—traits more adapted to the environment in which the individuals live.

MUTATIONS

The variation covered to this point is due to the shuffling of gene variations already present within a population and their varied effects on traits. Other sources of variation involve mutations of the genetic material—alterations of the DNA that occur daily in every species on the planet. The key importance of mutations is that they can generate even more variation in the form of new alleles and genes, and as such are a key factor in allowing continued evolution. In most one-celled prokaryotes, any mutation to the genome (that is not fatal) is potentially heritable since they only have one set of genes to pass on to the next generation through their asexual mode of reproduction. In eukaryotic protists (those that are unicellular), most mutations are likewise potentially heritable since a good deal of their reproduction is accomplished by asexual fission, or by the creation of gametes from that one cell's genome. In most multicellular eukaryotes, a mutation must typically occur in germ line cells (those that give rise to eggs or sperm) to be potentially heritable, and in large multicellular organisms the majority of genetic mutations would be occurring in the far more numerous somatic cells (non-germ line body cells) where they might cause cell death, cancer, have minor effects, or have no effect at all. Whatever the outcome of somatic mutations, they will not normally be passed on to offspring unless the organism engages in asexual reproduction that involves somatic tissues, and this does occur in some kinds of plants and simple animals.

If new mutations are passed on to offspring, the new alleles and genes then become part of the variation that natural selection will screen in that generation. The great majority of mutations are typically either harmful or neutral in their effect on the organism. Mutations are very often harmful because most of the important coding genes were already working pretty well in organisms (the parents) that had survived to reproductive age. Also, they were obviously inherited from a line of fit ancestors. Any change in genes with such a proven track record is likely to make those genes work less effectively—or not at all. A smaller percentage of mutations will be neutral, as is discussed further in Chapter 8. An even smaller number of mutations prove to be advantageous to the organism to some degree, and so have a better chance of surviving and even spreading in the population.

We have come to understand that at least some of the noncoding DNA (DNA that does not directly code for proteins or known products) is important in regulating the coding genes. Regulation involves turning coding genes on and off as well as determining the timing of when they become active—and for how long. Variation in these regulatory DNA sequences is also important in generating variations in individuals that natural selection can work on. The study of gene regulation has grown into a huge, diverse, and complex sub-discipline of molecular biology—most of it far beyond the scope of a general book of this type.

Point Mutations

Mutations can involve almost any fraction of the genome, all the way from single nucleotides up to the whole genome. Point mutations involving single nucleotide changes in the DNA are the most common type, and the three basic types of point mutations are substitutions, deletions, and additions. The genetic code is presented in Chapter 8 in Figure 8.1. Living organisms code for the construction of proteins using this language of 64 three-base (or three-nucleotide) "words" known as codons constructed from the four nitrogenous bases found in DNA. Those four bases are adenine (A), thymine (T), guanine (G), and cytosine (C). Though many of the codons are synonyms coding for the same amino acid (explained in Chapter 8—and obvious in Figure 8.1), most of the different codons code for different amino acids and their respective placement in the protein.

A single substitution point mutation affects only one base in one codon. It might, for example, change the codon T C A to T C G. Looking at the genetic code, we find that this would be a neutral mutation since both T C A and T C G code for the amino acid serine (Ser). The protein produced from the gene containing this mutation would be unchanged from that of the previous "normal" gene. However, if the third base was changed to a T (T C T), the codon would then code for the amino acid arginine (Arg). If arginine is coded into the protein where serine should have gone, we would have a slightly different protein— one that might not function at all, or function poorly, or be essentially equal to the normal protein, or possibly work somewhat better than the previous form of the protein. Substitutions are fairly common compared to most other mutations, and they are a major source of new alleles since they often create another variation of the affected gene.

The other two forms of point mutations are deletions and additions, where one base is deleted from the DNA— or one base is added to the DNA. Both of these types of point mutation are typically disastrous, usually producing a completely malfunctioning gene. The reason for this is because the genetic code is read in threes—three bases in a row forming a codon that codes for one amino acid. Deleting one base will cause all bases downstream of the mutation to be read in a shifted reading frame, which will generate a nonsense protein with almost no chance of being adequately functional (Figure 5.2). In Figure 5.2, the part of the normal gene shown would have coded for the chain of amino acids: methionine–valine–serine–proline–leucine–alanine. You can check this out using Figure 8.1 in Chapter 8 if you want to take the time. If the fourth A is deleted as in the second part of Figure 5.2, and the reading frame from that point on is shifted to the right by one base, the gene now codes for methionine–valine–serine–leucine–tryptophan–? Finally, if the base G is added as indicated in the third part of Figure 5.2, the amino acids coded for would be methionine–valine–glutamine–serine–serine–glycine–? As you can see, deletions or additions alter the coding of most of the downstream amino acids. This would typically create a nonsense protein having no utility in the cell. Here we have the "garbage in–garbage out" principle, with deletions and additions essentially turning a precise code into garbage.

Chromosomal Mutations

Larger scale mutations involving significant portions of one or more chromosomes are known collectively as chromosomal mutations, of which there are several types. These too have had profound effects on the evolution of life. One kind of chromosomal mutation is a *deletion*, where a section of a chromosome is deleted and lost from that

Leading section of a normal gene—which would continue to the right

T A C C A A T C A G G A G A C C G T →

T A C C A A T C*G G A G A C C G T →

A deletion of the fourth adenine (A) in the original gene, position marked by the *

T A C C A A G T C A G G A G A C C G T →

An addition of an extra G (underlined) into the original gene

FIGURE 5.2   Point mutations. Illustrating first the deletion of a single base, and then the addition of a base into the original gene. See text for further clarification and examples.

chromosome—usually during a cell division event. If the segment in question contained essential genes or regulatory elements, any cell receiving the deficient chromosome would likely die if it needed those genes, and certainly any sexually produced offspring receiving this deficiency would not survive or develop.

Another chromosomal mutation is a *duplication*, in which a section of a chromosome is duplicated and reinserted somewhere along the same chromosome such that it now contains some duplicated material, likely including some duplicated genes. If either a deleted section from one chromosome or a duplicated section of a chromosome happens to be inserted on another non-homologous chromosome, a *translocation* has occurred—material having been "translocated" to that other chromosome.

Yet another possible chromosome mutation is the *inversion*, where a section of a chromosome is split out and reinserted in reverse order from the original condition. Assuming the cuts and insertions do not occur within important coding or regulatory regions, the cell, organism, or even offspring may experience no harmful effect from an inversion. Figure 5.3 illustrates these major categories of chromosomal mutations.

Less common chromosomal mutations include *fusions* and *fissions*. These are just what they sound like—two chromosomes fusing into one larger chromosome, or one chromosome "breaking" (fission) into two smaller chromosomes. Perhaps the best-known example of a chromosomal fusion occurred in our own human line of ancestry sometime after our line split from that of chimpanzees and bonobos—roughly a little over six million years ago. The latter two apes (as well as gorillas and orangutans) have 48 chromosomes as their diploid number. Humans have only 46 because our so-called second chromosome (or pair two in the diploid state) is actually the result of the fusing of what were two separate chromosomes back in our ape ancestral tree. Genomic scientists have worked this out in great detail and have pinpointed exactly where this fusion took place within our chromosome #2.

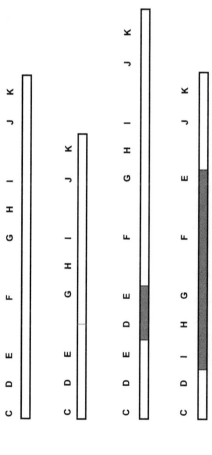

FIGURE 5.3   Some possible chromosome mutations. The first bar represents a stretch of a "normal" chromosome with several genes along its length indicated by the letters C through K. The second bar represents the same chromosome section following a *deletion* of a section of the chromosome containing gene F; therefore F is now missing from this somewhat shortened chromosome. The third bar represents a *duplication* of the section of the chromosome containing genes D and E. In this representation the duplicated section is inserted adjacent to the original and identical section. The last bar represents an *inversion* of a sizable section of the chromosome that contained the genes E through I. Though not illustrated, if a deleted or duplicated section of a chromosome inserts into a different non-homologous chromosome, the mutations would be known as a *translocation*, and like the duplication should make the receiving chromosome a bit longer. (For a color version of this figure, the reader is referred to the online version of this book.)

Chromosomal fission, the splitting of one chromosome into two or more smaller chromosomes, is also believed to have contributed to the "genomic architecture" of some species and clades. Even across distinct populations of the same species or among closely related species in the same genus, there are examples of small differences in chromosome number that are believed to be the result of fusions, fissions, or both.

Polyploidy can be considered to be the largest scale chromosomal mutation in that all the chromosomes are replicated three or more times in the resulting offspring. Polyploidy is a rather complex topic that is discussed in more detail in Chapter 14, so for now I would simply point out that while polyploidy is an extremely rare event compared to most other chromosomal mutations, over geologic time it has occurred a great many times, and it has left its record in the genomes of many modern species and clades—especially so in several plant groups. Importantly, it can lead to rapid speciation events and a diversity of phenotypic changes in the new polyploidy individuals.

Mutations from Transposable Elements

Transposable elements will be the major topic of Chapter 6. Active transposable elements can insert themselves almost anywhere along a chromosome, so of course they too are a source of genomic mutations. If they insert within exons, the effect will be significant and far more often than not disastrous. If they insert in regulatory genes or promoter regions of DNA, they will have unpredictable effects, though one would bet on a harmful outcome for any one insertion. Still, over millions of generations in millions of species at least some of these insertions of additional DNA have proven either neutral or actually beneficial in some way. Again, more on this in the next chapter, but a final consensus as to the overall role of transposable elements in genomes and evolution is still a distant goal.

It has been recently discovered that far more than the 1.5% of coding DNA is transcribed into RNA, and at least some of this noncoding RNA is now known to have regulatory functions in cells—involved in yet another level of cellular regulation. MicroRNAs are one class of transcribed RNA that is now known to have regulatory functions—usually in blocking particular coding genes from being active. Current estimates of the percentage of the genome involved in important regulatory function range from 20% to 80%. Hopefully this very important question will be elucidated more completely in the near future. This point has of course great bearing on how much of the genome can be affected in one way or another by mutations.

Epigenetics

All the foregoing discussions in this chapter have covered mechanisms that potentially altered the actual letters (bases) of the genetic code—or the structure (architecture) of the genome. Over the past few decades, a different set of mechanisms in gene alteration has been elucidated and termed *epigenetics*. Epigenetics refers to a form of gene regulation that most often involves "tags" consisting of other molecules attaching to genes and/or associated chromosome proteins that either turn genes off, or on. One common type of epigenetic tag consists of a methyl group (a carbon atom with three attached hydrogens) which can be attached to one or more of the cytosine bases in the genetic code (the C in the A, T, C, G gene alphabet). This methylation is actually just the first step in a cascade of molecular events that results in a shutdown of the affected gene and possibly the condensation of that area of the chromosome into heterochromatin (a more tightly coiled and inactive stretch of a chromosome). Less condensed and more active DNA (actively transcribing genes) is termed euchromatin. Another type of epigenetic tag consists of an acetyl group that bonds to a part of a histone molecule. DNA in normal chromosomes is actually wound around many globular histone complexes—almost two turns of DNA around each histone unit. You might for analogy imagine a string of pearls but with the string wrapped around each pearl rather than passing through it. As opposed to methylation, the "acetylation" of a histone tends to turn an associated gene on. Yet a third set of epigenetic mechanisms involves small RNA molecules that are produced to block the translation of other mRNA molecules. These are referred to as interfering RNA molecules because they interfere with the "normal" translation of mRNA—which would otherwise result in the production of proteins. And there are yet a few more epigenetic mechanisms that have been elucidated, and probably still more await discovery. In short, epigenetics appears to be a very complex layer of gene regulation in addition to that already known, which involves specific DNA segments and modifier proteins. It is now believed that epigenetic mechanisms are responsible for much of the cell and tissue uniqueness that emerges in the development of multicellular organisms. That is, liver cells divide and form more liver cells, while skin cells divide and form more skin cells. Each cell type is of course using only a small fraction of the total genome that every cell inherits from the zygote (fertilized egg) from which all the organism's cells arise. Mammals contain more than 100 specialized tissue types, with each type using a unique fraction of the total genome. Many genes used by liver cells are not used by skin cells—and vice versa. The genes that are turned off in specialized cells are typically turned off (and on in some cases) by epigenetic mechanisms. Broadly stated, epigenetics in this usage refers to situations where genetically identical cells grow and "behave" differently due to "other" molecular and environmental influences.

The interesting point is that when the chromosomes divide before a cell division to form two new skin or liver cells, the epigenetic tags are replicated along with the chromosomes and their genes, as are the activities of certain interfering RNAs. So what does this have to do with evolution? For one thing, many organisms such as amoeba, some plants, and even some animals reproduce asexually through mitotic cell divisions similar to those occurring in the tissues of multicellular organisms, so asexually produced offspring can inherit the epigenetic tags and other mechanisms just like they do in the liver or skin cell examples. The most unexpected finding about epigenetic gene modifiers is that they not only get replicated as cells divide mitotically, but some (especially the methylation tags) are now known to be capable of being copied into and transmitted by germ line cells—the egg and sperm of sexually reproducing organisms (produced typically by meiosis) such that these modifiers are heritable by the next generation. So, epigenetic inheritance occurs when epigenetic modifications to the DNA are passed from one generation to the next, and a few examples have surfaced in recent years.

One of the best examples of epigenetic inheritance was found in a plant called toadflax (*Linaria vulgaris*). Going all the way back to Linnaeus, two varieties were recognized and assumed to be different species. In one form, the flowers were radially symmetrical, but in the other they were bilaterally symmetrical—an obvious phenotypic distinction. Only in recent years was it discovered that the genes responsible for flower symmetry are identical in these two varieties, but their epigenetic tags differ (Jablonka and Lamb, 2005). This epigenetic inheritance has undoubtedly been passed on for a great many generations of toadflax, though it is still a mystery as to what circumstances or influences caused the initial divergence of flower symmetry through epigenetic modification.

A few examples of epigenetic inheritance have been documented in animals, but often with the inheritance of the epigenetically modified trait not being as reliable or predictable as it appears to be in the toadflax example, and typically not for as many generations in a row. The major point in all of this is that some epigenetic modifications appear to be the direct result of environmental factors such as temperature or diet. In short, this seems like a case of Lamarckian evolution—acquired characteristics being inherited by offspring. This of course is something that was rejected by most evolutionists for well over a hundred years, due to the belief that all inherited traits were the result of the precise base coding of the DNA in genes.

Another finding has been that some epigenetic inheritance effects are dependent on the sex of the parent a gene/allele was inherited from. Some alleles, for example, will be expressed if inherited from the male parent, but the same allele if inherited from the female parent would be silenced by methylation tags (or vice versa in some cases). This parent-dependent effect has been given the name of "genomic imprinting", and a number of examples have been documented. Whatever the mechanism or cause of epigenetic effects, it must be remembered that these mechanisms require the assistance of proteins and enzymes coded for by the genome, so they really are just phenotypic options already "programmed into" the genome, but awaiting a particular stimulus to take effect.

It is still too early in our understanding of epigenetics (and the known examples too few) to say exactly how important these phenomena are to the long-term evolution of populations over hundreds of generations, but it is a question that is being asked and pursued by many scientists today. As of now, there are simply too few long-term studies covering several sequential generations to establish the intergenerational stability of epigenetic modifications. Also, many of the known trait effects associated with epigenetic modifications and inheritance are not obviously of adaptive value. Still, if it turns out that epigenetic inheritance can be reliably inherited for several generations and provides some advantage, it may act as a buffer in some instances, allowing a needed phenotypic modification to carry the population through a period of time long enough for new mutations to occur that might better suit the organisms for survival more directly (without the continued need of the respective epigenetic modifiers). As our understanding of evolution and its associated mechanisms and processes improves, we sometimes have to reformulate the structure of that understanding and reassess the importance of some mechanisms—thus: evolutionary biology evolves.

Chapter 6

Transposable Elements, Viruses, and Genomes

The importance of transposable elements as generators of mutational variability comes from the fact that they produce mutations that are unlikely to arise by any other means.

Austin Burt and Robert Trivers

It seems likely that the utility of TE insertions is merely a rare side effect rather than the reason for their existence. Most insertions are probably disadvantageous, end up losing their ability to transpose, and then slowly dissolve into "junk" DNA.

Andrew Pomiankowski

If mutations, introns, gene duplications, pseudogenes, vestigial genes, and gene losses are not collectively complicated enough, these topics still do not cover all the complexity and strangeness to be found within genomes. Another significant component of many genomes consists of what are known as transposable elements. These appear to be fundamentally DNA parasites, most of which were incorporated into various eukaryote lines a very long time ago, though many continue to proliferate within and between some lines of descent (Avise, 2010). A few transposable elements have been found in prokaryotes, but they are far more common in the domain of the eukaryotes.

One usually thinks of infection and parasitism in terms of an individual who becomes infected, then gets sick, and then usually becomes well again—having eliminated the virus, bacteria, or other disease agent from the body due to

Evolution. http://dx.doi.org/10.1016/B978-0-12-800348-0.00006-7
Copyright © 2014 Elsevier Inc. All rights reserved.

immune mechanisms and/or medical intervention—but not so with transposable elements. They are parasites that have existed in various lines of evolutionary descent for literally millions of generations—meaning that you inherited them from your parents, you will have them your whole life, you will pass them on to your offspring, and there is no way to rid them from your cells. One hypothesis on the origin of transposable elements is that they originated as viruses which incorporated their DNA into host genomes and subsequently lost the ability to escape the host in protein capsules—as most "normal" viruses tend to do. This is a very deep and complex topic that we need not enter into too exhaustively here. Suffice to say that transposable elements were most probably aliens originally—some now existing as unnecessary DNA "passengers" riding down through time in host genomes. They are wonderful examples of the selfish drive of DNA to get itself copied and passed on through the generations. They also give weight to the argument that "genes" (or any DNA message that can be copied into offspring) are the units of selection—discussed briefly in Chapter 2.

The classification of transposable elements is complicated, but the two major divisions are *transposons* and *retroelements* (also called retrotransposons). Functional transposons are stretches of DNA that can cut themselves out of the chromosome they are in and move to other places within the genome, basically a cut and paste process. They average about 2,500 base pairs in length and make up just under 3% of the human genome, but even at that they exist in about 300,000 copies (Fairbanks, 2007). Actually, all human transposons have long ago mutated such that they no longer possess their cut and paste abilities. In some species, some transposons are still able to move about within the genome. They were earlier termed "jumping genes" after Barbara McClintock discovered them back in the late 1940s. Apparently they were still active in certain species of "Indian corn" where they affected the color of the seeds (the corn kernels). McClintock was studying corn genetics and inheritance, and so stumbled onto this weird group of DNA entities.

One transposon known as Tc-1/mariner is found in several animal phyla, including our own phylum—the Chordates. Humans have about 14,000 copies of this one transposon alone, and since it is shared by all chordates and even insects, it has undoubtedly been around for a very long time indeed—having been inherited through common descent from the time when the animal phyla were just diversifying, some 600 million years ago (Fairbanks, 2007).

The other major category of transposable elements is the retroelement. These DNA regions have the unique property of being able to have themselves transcribed into mRNA, and then "retrocopied" back into DNA as a copy of the original retroelement. This copy is then inserted "somewhere" back into the genome. In short, this is a copy and paste process that allows retroelements to actually multiply within genomes. They span a range from 300 up to about 8,000 base pairs in length. Thanks to their ability to replicate, they typically represent a more sizable percentage of individual genomes. In humans they account for a whopping 43% of the human genome—existing in about 2.7 million copies divided into a few major categories (Fairbanks, 2007).

Like our transposons, most of these too have long ago mutated such that they are now unable to perform the copy and paste function, so they are just getting passed on "passively" from generation to generation, but there are a few retroelements that are still viable and active. One of these still-functional retroelements is called *Alu*. It is only about 300 base pairs long, and it is the most common retroelement in primates. In humans, Alu exists in more than 1 million copies, and these collectively make up about 12% of the human genome (Fontdevila, 2011). You might consider again that the coding gene exons represent only about 2% of the genome! Transposable elements likely account for even more than 50% of the human genome because really ancient copies will have mutated to the extent that we can no longer recognize them as such—their nature and relationships having been masked by multiple mutations over time.

Going with the definition of evolution as change in genomes over time, it is clear that transposable elements have made sizable contributions to the evolution of a significant number of lines of descent, again especially so in the eukaryotes. What we see today are the many copies of these elements that have been added (long ago) in places that did not significantly cripple the genome and that are now either passive riders through time, or in some cases actually serve a functional role in their respective genomes. Undoubtedly a huge number of past insertions of both transposons and retroelements had disastrous and fatal results when they affected exons and regulatory DNA regions, so we have no record of those events. But by slowly "bloating" the noncoding portions of DNA, they helped promote the survival of even more transposable element insertions, which now have many more relatively harmless places at which to insert.

Again, like all mutations, most insertions of transposable elements have undoubtedly been harmful or neutral. Inserting 300 to 8,000 continuous base pairs randomly into the genome is a far more significant mutation than the simple substitution of one base for another—believed to be the most common type of point mutation. Whereas a significant percentage of these point substitutions are neutral or nearly so, this is unlikely to be the case with transposable element insertions into coding regions. Some surviving harmful insertions of transposable elements are now known to be the cause of some genetic diseases and cancers, and more examples of these harmful effects will likely be discovered in the future.

Still, it is possible, and even probable, that some of the genomic change due to the past activity of transposable elements has been beneficial. A few genes are now known to have transposable elements as part of their

coding exons, and a few control regions of DNA likewise show evidence of transposable element insertions
that apparently tweaked the activity of the target genes
in a neutral or beneficial way. That is only to be expected
given the vastness of the genome and the exceedingly
long history of insertion activity that genomic studies
have started to reveal. The first lead quote of this chapter is from an excellent book (Burt and Trivers, 2006,
Genes in Conflict) on selfish genetic elements, which
details more widely and specifically the complexity
of transposable elements, including specific examples
of some possible beneficial outcomes. Clearly, these
selfish DNA elements have had significant effect in a
great many lines of descent, and they are therefore a
major force that has shaped—and continues to shape—
genomic evolution.

There are also a significant number of genetic elements that appear to be more clearly of recent viral origin. These endogenous retroviral units of DNA seem to be
the remains of retroviruses that reverse copied their DNA
into the human nuclear DNA a great many times (typically many copies of many various retroviruses). These too
are now mutated into functionless and silent stretches of
DNA, but they make up about 8% of the human genome
(Avise, 2010). Like the transposable elements, they may be
mostly useless parasites except for a few that have located
in the genome in a way that provides a useful function.
As a cautionary note, and as also mentioned in Chapter 8
that deals with neutral evolution, recent work is suggesting
that a significant fraction of the noncoding DNA (including transposable elements) performs assorted regulatory
functions in eukaryotic genomes. This topic is complex
and will certainly take more work and time before we have
a reliable answer as to how much of the genome has important functionality, and what fraction might still validly be
termed "junk DNA". Indeed, it seems that much more than
half of our nongene DNA transcribes RNA molecules, a
few of which are now known to have functionality in gene

regulation. But transcription of RNA alone does not necessarily imply functionality. This question of functionality (in this case how much of the genome is functionally important) constitutes one of the major questions left in gaining a clear understanding of genomes.

Chapter 7

Horizontal Gene Transfer

In my view, it is no longer a matter of sensible dispute that HGT is a defining process in the evolution of prokaryotes that affects all aspects of bacterial and archaeal biology.

Eugene V. Koonin

Conjugative or mobilized plasmids are infectious genetic elements that can spread throughout entire populations and subspecies and even to different species and Kingdoms.

Gunther Koraimann

Let me start this chapter by emphasizing again that typically it is the heritable genetic material of *some* individuals that is copied and passed from one generation to another, and it is therefore the total pool of these genetic entities that evolves. Though evolutionary change can occur over several generations due to genetic variation already present in a gene pool, really dramatic evolution over longer time spans typically involves the appearance of new genes and alleles. Traditionally, the thinking was that new genes and alleles only come into existence through various mutations, as discussed in Chapter 5, but they can also appear in species through a process termed *horizontal gene transfer* (HGT)—a process that has been known for some time, though the importance and range of its role in evolution are only currently being fully realized and appreciated. The term horizontal here contrasts this mechanism with the more typical vertical transmission of genes "down" to descendants as typically occurs in eukaryotes through the reproduction of offspring. In HGT, genes or larger

Evolution. http://dx.doi.org/10.1016/B978-0-12-800348-0.00007-9
Copyright © 2014 Elsevier Inc. All rights reserved.

DNA units can be moved from one "adult" prokaryote cell into another that is not an offspring of the donor cell. This would be like you moving some of your DNA into another living person in your environment—something only recently possible through very advanced technology, but which essentially does not occur at all in humans or other mammals. Even more amazing is that in prokaryotes, HGT sometimes occurs between very different bacterial and archaeal species. This would be like you transferring some of your DNA into a tiger, an earthworm, or even an apple tree!

HGT (also known as *lateral* gene transfer) is relatively common among prokaryotes—the bacteria and archaea— and the more common mechanism for HGT is known as *conjugation*. Prokaryotic cells typically contain one circular chromosome that bears the essential genes plus the required regulatory DNA for that cell. Some cells in the population can contain an additional and much smaller circle of DNA known as a plasmid. Plasmids "behave" as though they "want" to get copied and transferred into other cells, since their genes (plasmids contain a few genes) direct the formation of straw-like structures called pili (singular—pilus) on the surface of their host cell. If possible, a pilus will attach its distal end to another bacterial or archaeal cell. The pilus then shortens, bringing the two cells into contact, at which point a cytoplasmic bridge is formed between the two cells (Figure 7.1). As all this is happening, the plasmid in the donor cell copies itself so that a new copy of the plasmid can be transferred through the narrow cytoplasmic bridge into the receiving cell. Donor pilus-forming cells always contain the plasmids, while receiving cells lack the plasmids. This transmission of plasmids looks for all the world like a bit of parasitic DNA copying itself *to infect* yet another uninfected cell. The cells lacking plasmids are generally perfectly normal and capable cells, and receiving a plasmid in most cases is of no benefit to the now infected cell. Plasmid DNA is known to occasionally end up being incorporated into the larger circular chromosome

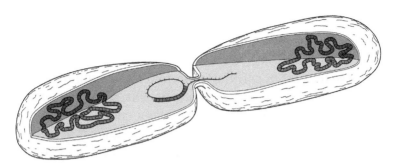

FIGURE 7.1   Bacterial conjugation. The bacterial cell on the left contained a plasmid. It earlier formed a pilus to attach to the bacterial cell on the right lacking its plasmid. The two have now been drawn together and a cytoplasmic bridge has formed connecting the cytoplasm of the two cells. The plasmid is here shown in the process of ejecting one strand of its double helix DNA, which is now moving into the cell on the right lacking the plasmid. After this transfer is complete, both plasmid strands will build the missing half to form complete double stranded plasmids in both cells. The two cells will then disconnect and go their separate ways. *Figure by Jeff Dixon.*

of the cell, so this process definitely does contribute to the evolution of prokaryotes—whatever the interpretation of these transfer events.

Much has been made of the ability of HGT to confer adaptive traits on the receiving cells—most commonly mentioned are resistance to some antibiotics, additional virulence traits that allow the organism to become a more virulent pathogen, or increased metabolic abilities. However, given the many billions of times this process occurs and the few known benefits associated with this phenomenon, it seems reasonable to assume that most of this plasmid "infection" is either neutral or slightly harmful to these cells—really amounting in the great majority of cases to a form of mutation or parasitism. In the rare event that the plasmid does confer adaptive benefit (adaptive in the current environment), then of course the few deriving that benefit will quickly outcompete and outgrow the competition—just like the case of the rare beneficial mutation. Remembering that evolution cannot plan for the future or carry harmful mechanisms forward because they "might" benefit future individuals would argue against HGT as being an innately beneficial process for the individual cells involved. Some suggest that it is "allowed" to occur because the cost of carrying a plasmid may be minimal and can in rare situations confer some benefit, but this is a hard point to prove or even interpret without question.

A somewhat rarer method of HGT is known as *transduction*. In transduction, phages (viruses of prokaryotes) actively reproducing in a host cell in some instances package some of the host DNA in addition to their own replicated viral DNA. When these viruses then escape to infect other cells they can "deliver" that host DNA into other cells of either the same or of a different species. Questions remain as to how common and important transduction is as compared to conjugation, though we do know that bacteria and archaea are continually plagued by phages.

A third mechanism termed *transformation* is also known to occur. In transformation, some cells simply "take up" DNA that exists in their environment. Since cells are always dying and being lysed in most microbial environments, DNA is commonly present in those extracellular fluids. Prokaryotes that can take in this environmental DNA are said to be "competent", and there are various hypotheses for the reasons why this might benefit such cells (use of the DNA as nutrients, as aids to DNA repair, as sources of genetic variability, etc.). Most transformation appears to involve the intake of DNA that is homologous with some of the DNA inside the receiving cell, so this would typically be derived from dying conspecific cells in the environment. Importantly, the piece of new DNA taken in is often switched into the chromosome replacing the original homologous section. This method too has the potential to introduce genetic variation into cells. When intraspecific homologous DNA is so incorporated, it really is somewhat like a form of indirect sex.

In all methods of HGT, at least some of the DNA being acquired is coming from different species of prokaryotes. In nature, prokaryotes typically live in diverse communities of several different species—whether it be in biofilms, in the soil, in aquatic environments, or in the guts of animals. The opportunity to acquire unique DNA from other species is ever-present due to their environments and to the various mechanisms of HGT available to them.

Yet another mechanism of HGT occurs in the case of endosymbiotic unions, as occurred in the mergers that gave rise to the mitochondria and chloroplasts of eukaryotes (Chapter 12). In these mergers, the "engulfed" microbe eventually loses some unnecessary components of its genome, but also transfers many of its genes into the genome of the host cell. This is a far rarer event than any of those mentioned above, but the outcome of such events can be revolutionary in terms of their eventual evolutionary outcomes—as in the origin of the Eukaryota domain.

It is hard to quantify the amount of evolutionary change accounted for by HGT, but since we suspect that this process was going on long before eukaryotes even existed, HGT has undoubtedly occurred an astronomical number of times over at least the last three billion years—a time span much longer than many of the more commonly recognized organisms like ants, whales, or birds have existed. Since we define evolution in terms of genomic changes within populations over time, most of evolution has actually occurred in prokaryotes (contrary to popular conceptions). Since eukaryotes evolved in part from symbiotic mergers of prokaryotic cells (Chapter 12), much of the early evolution of our lines as prokaryotes (originally) was undoubtedly shaped by untold numbers of HGT events billions of years ago.

A newer significant discovery concerning HGT is that some is strongly suspected to have occurred between eukaryotes and bacteria or archaea, again with viruses being one possible transfer mechanism. There are now many clear cases of gene transfer from bacteria to eukaryotic protists that live in the same environments; such as aquatic environments, the guts of some animals, and bottom muds (Andersson, 2006). Almost all multicellular eukaryotes have symbiotic prokaryotes on or inside them in great abundance and diversity—as in the guts of most animals. Here, DNA is regularly released from dying host eukaryote cells and symbiont prokaryote cells in close proximity to healthy cells of both, possibly allowing for something like transformation to potentially occur. Also, most insects and nematodes use intracellular digestion in addition to extracellular digestion. In intracellular digestion, small bits of solids along with some of the lumen liquid are taken into food vacuoles by cells lining the intestine. Here is perhaps another avenue for the entrance of DNA into animals, though the normal expectation would be that this incorporated DNA would be digested within the food vacuole. Some nematodes that are parasitic on plants appear to have acquired cellulase genes from either bacteria or fungi, though the mechanism of this acquisition is still in question (Danchin, 2011).

There are even some possible cases of eukaryote-to-eukaryote HGT. One example appears to be the presence of very similar "endogenous antifreeze proteins" in three distantly related species of cold water marine fish. Both the coding regions and the intron structure of the gene for this antifreeze protein show significant similarity in all three species, but the gene is absent in the close relatives of all three—seeming to negate vertical gene transfer through lines of descent as the explanation (Graham et al., 2008). Perhaps this gene was transferred by viruses that affect at least two of these species. Another proposed hypothesis is that free DNA released from the dying eggs or sperm of one species might be taken in by the viable sperm or eggs of other species spawning at the same time, thereby effecting an HGT event. Undoubtedly more cases and potential mechanisms will be forthcoming as biologists elucidate and compare more eukaryotic genomes.

In the big picture of evolutionary history, HGT makes much of the tree of life into a network of gene movement rather than a straight-line branching affair as would be the case with vertical gene transfer only. Still, the majority of gene transfers over time are vertical, even in the pro-karyotes, and most of the genes of any bacterial species have been passed vertically within that species for at least millions of generations. The many diagrams being shown in textbooks and other sources emphasizing a network of genetic relationships are somewhat misleading since in the great majority of HGT cases, only a small amount of DNA is being shared or transferred. These are not mergers of two genomes, which these diagrams might seem to imply. Only in the far fewer cases of endosymbiosis (as in those lead-ing to mitochondria and chloroplasts) have whole genomes been combined (Chapter 12). Still, since bacteria and archaea are alive and well and extremely diverse—with an evolutionary history of at least 3.6 billion years—HGT has certainly been, and continues to be, an important contribu-tor to the evolution of life on Earth. With evidence now that at least some eukaryotes have experienced HGT events, HGT clearly qualifies as one of the major mechanisms of evolutionary change on our planet.

Chapter 8

Neutral Evolution

As a rule of thumb, only a small proportion of the thousands of letters in a gene are particularly important; the rest can vary more or less freely as mutations build up over time, because the changes don't matter much and so are not eliminated by selection.

Nick Lane

Perhaps the majority of alleles are "neutral," that is, their mutation does not affect the fitness of the phenotype. As a result, it is now realized that seemingly identical phenotypes may conceal considerable variation at the level of the gene.

Ernst Mayr

If we are to define evolution as any change over the generations in *the genomes* of populations and species, as was discussed in Chapter 1, much of the genomic change that occurs will be neutral in terms of effects on the phenotype and fitness of the organisms involved. We have come to understand that neutral genomic change in fact probably accounts for the majority of total genomic change that has occurred over evolutionary time. Some workers have argued that if a genomic change does not translate into any detectable alteration in the morphology or physiology of an organism, it really is not evolution, but this stance runs counter to the modern definition of evolution that focuses on genes or genomes—as opposed to traits (the phenotypes that result from *some of* the genetic material).

A short four or five decades ago, the only hard information available on the variations in populations was the obvious outward morphology and the protein differences

Evolution. http://dx.doi.org/10.1016/B978-0-12-800348-0.00008-0
Copyright © 2014 Elsevier Inc. All rights reserved.

that could be detected using immune techniques (like blood typing). Countless workers then began to learn and use the new (at that time) technique of gel electrophoresis to look for variations in protein structure (often in enzymes) within and among populations and species. A surprising outcome of this electrophoresis revolution was that far more variation was detected than had ever been suspected; in short, many more alleles were present than had been assumed. The notion at the time was that natural selection should eliminate all but the "one best version" of such important alleles as those coding for enzymes. The question then arose as to how all this newfound variation could exist within the paradigm of Darwinism. The answer we now know is that most of these alleles are neutral, in that they work essentially equally well in coding for an effective protein.

To explain neutral evolution and variation, we will enumerate and discuss briefly most of the ways in which changes in the DNA (the genome) might be neutral in effect.

THE REDUNDANT GENETIC CODE

We now know that the genetic code consists of all the three base combinations possible using the four DNA bases thymine (T), adenine (A), cytosine (C), and guanine (G). There are 64 such three-base combinations (known as triplets or "codons"), shown in Figure 8.1. In each of the four major columns, the first three-base groupings in capitals are the DNA codons present in the genes. The second three-base groupings in italic capitals are the messenger RNA codons that would be produced in the process of transcription. If you are unfamiliar or rusty concerning this process, a quick reference to a general biology text or an instructive website is recommended. The "U" in many of the RNA codons stands for uracil, which is a base found only in RNA; it is the functional equivalent of thymine (T) in DNA. Finally, the abbreviations following these codons are abbreviations for the amino acids that would be coded for by each respective

AAA *UUU* Phe	AGA *UCU* Ser	ATA *UAU* Tyr	ACA *UGU* Cys
AAG *UUC* Phe	AGG *UCC* Ser	ATG *UAC* Tyr	ACG *UGC* Cys
AAT *UUA* Leu	AGT *UCA* Ser	ATT *UAA* TM	ACT *UGA* TM
AAC *UUG* Leu	AGC *UCG* Ser	ATC *UAG* TM	ACC *UGG* Trp
GAA *CUU* Leu	GGA *CCU* Pro	GTA *CAU* His	GCA *CGU* Arg
GAG *CUC* Leu	GGG *CCC* Pro	GTG *CAC* His	GCG *CGC* Arg
GAT *CUA* Leu	GGT *CCA* Pro	GTT *CAA* Gln	GCT *CGA* Arg
GAC CUG Leu	GGC *CCG* Pro	GTC *CAG* Gln	GCC *CGG* Arg
TAA *AUU* Ile	TGA *ACU* Thr	TTA *AAU* Asn	TCA *AGU* Ser
TAG *AUC* Ile	TGG *ACC* Thr	TTG *AAC* Asn	TCG *AGC* Ser
TAT *AUA* Ile	TGT *ACA* Thr	TTT *AAA* Lys	TCT *AGA* Arg
TAC *AUG* <u>Met</u>	TGC *ACG* Thr	TTC *AAG* Lys	TCC *AGG* Arg
CAA *GUU* Val	CGA *GCU* Ala	CTA *GAU* Asp	CCA *GGU* Gly
CAG *GUC* Val	CGG *GCC* Ala	CTG *GAC* Asp	CCG *GGC* Gly
CAT *GUA* Val	CGT *GCA* Ala	CTT *GAA* Glu	CCT *GGA* Gly
CAC *GUG* Val	CGC *GCG* Ala	CTC *GAG* Glu	CCC *GGG* Gly

FIGURE 8.1   **The Near-Universal Genetic Code: how DNA codes for protein construction—see text for clarification.**

codon in the process of translation. Three of the codons (UAA, UGA, and UAG) serve as terminator codons that signal the end of the message, but do not code for amino acids. Note that only the 20 amino acids found in most organisms need to be coded for, so we have many more DNA codons than amino acids that need to be coded for.

The evolutionary "solution" for this discrepancy was to allow most of the extra codons to be synonymous with others, such that two, three, four, or even six different codons might all code for the same amino acid. Looking at the chart, you can see that a DNA change from CAA to CAG would still code for the same amino acid—valine (Val). Most of the synonyms involve a difference in only the third base in the triplet, but also note that AAC and GAG are synonymous—with both coding for the amino acid leucine (Leu). This "redundancy" means that within the code there are many possible base substitutions (through point mutations) that would result in no change in the amino acid structure of the protein. When a base change does not affect the amino acid structure of the protein, it is called a neutral silent mutation—neutral because it had no affect on traits, and silent because it does not even affect the protein structure. Neutral silent mutations in fact cannot be detected by gel electrophoresis methods, since they only reveal differences in the amino acid structure of the proteins. Some insightful workers in those early days of gel electrophoresis noted this possibility—that some alleles could differ in their respective codon synonyms and yet go undetected. Today with the capability of rapid DNA sequencing, even these silent mutations can now be elucidated.

If it was found that in two species of snake from the same genus, all the members of one species had CGA in a certain position in a certain gene, while the members of the other species consistently had the synonym CGG, this difference would be one part of the evolved genomic differences separating the two snake species, even though the effect of the difference was neutral with respect to the protein being produced. Only in recent years has this ability to see silent neutral variations between species and higher taxa produced additional characters for phylogenetic determinations and discriminations.

NEUTRAL NONSILENT MUTATIONS

Some base changes resulting from mutations obviously do result in a different amino acid being coded for within the structure of a protein, but it turns out that some amino acid substitutions do not affect the performance of the enzyme or other protein in any measurable way. A few of the amino acids are similar in size, shape, and charge, and so can be changed one for the other without destroying the functionality of the protein. In the case of enzymes, if the amino acid substitution occurs in a part of the protein that does not alter the active site of the enzyme, the altered enzyme might work just as well as the standard model. Usually only a small part of the surface of an enzyme constitutes its active site—the place where it fits and interacts with its substrate molecules to cause a chemical reaction.

Such mutations would be expected to be neutral since they do not affect the workings of the protein, but they are not said to be silent since they do alter the amino acid makeup of the protein. There are situations in which a "neutral" nonsilent mutation could vary in fitness from the original gene form. One is where one of the two alternate amino acids was in short supply due to limitations in diet or metabolic abilities. Another would be where one of the alternate amino acids had to be synthesized by the organism while the other was readily available in the diet. Yet another case is where both had to be synthesized but one was more energetically expensive to make than the other. In any of these cases, one amino acid might have a slight fitness advantage over the alternate one, even though the resulting protein in either case was equivalent.

Though it would be impossible and perhaps unethical to perform the experiment, if we could somehow replace all of a person's cytochrome C with the cytochrome C from a rhesus monkey, the metabolism of the human would most likely be unaffected, since all indications are that the single amino acid difference in cytochrome C between humans and monkeys (out of a total 104 amino acids) is a neutral one in terms of metabolic function.

Some neutral nonsilent alleles do result in obvious phenotypic traits that can vary within a population. Just look at several human faces in a crowd for some obvious examples. Though actually shaped by the influence of more than one gene, note the variety of nose shapes that humans display, most of which have little or no bearing on a person's ability to survive or reproduce. Most humans have earlobes that can be classified as either attached or unattached. Most people have the unattached variation, with the earlobe hanging free below the ear's attachment to the head. A few people instead have their earlobe fused with the head/neck and lacking the "hanging" lobe. Perhaps attached earlobes are problematic for people who are into earrings, but other than that there is no indication that having one form or the other is in any way beneficial or detrimental to one's health or genetic fitness. What person would examine and judge a perspective mate in terms of their earlobes? Such neutral variation within a population is common, and it also exists between related populations and species. The genetic mutations that gave birth to these neutral traits are therefore another example and level of neutral evolution.

MUTATIONS IN INTRONS

It is now known that even between the start and end of a DNA gene, there are typically found one to several stretches of DNA that do not code for any part of the finished protein. This is especially common in eukaryotes. Though these stretches are typically transcribed into a mRNA (messenger RNA) right along with the exons (those stretches that do code for amino acid sequences), their corresponding RNA is clipped or edited out and the remaining exon mRNA coding fragments are stitched back together to form the "edited mRNA", which leaves the nucleus to be translated into protein molecules. In a crude analogy, some words in this sentence clearly do not fit: Ask not what your country can do for you, tables turned and black were they ask what you can do for your country. If the "tables turned and black were they" is edited out, we have a meaningful and recognizable sentence. In genetic language, this meaningless insert is an intron. Again, the meaningful parts that are retained in the final edited mRNA are known as the exons. Why eukaryotic genomes have so many introns is still not clearly understood. Introns have now been identified in some prokaryotes as well. In eukaryotes, introns often account for 80% or more of average total gene length and up to 20% or more of the total genome. The point is that mutations within introns will most likely have no effect on the gene product, since their transcribed mRNA is edited out before the protein is made.

MUTATIONS IN OTHER NONCODING DNA

A major discovery of the past two decades is that many eukaryotic genomes contain vastly more noncoding DNA than coding DNA, with the typical figure for humans being more than 97% noncoding. Much of this noncoding DNA seems to exist in the form of various genetic parasites such as transposons and retroelements (Chapter 6). These parasites are like internal viruses, some of which can copy themselves within the genome and then be passed on to offspring, but other than this single activity, they typically have little effect on the host or its genome except for a slight metabolic drain each time all this parasitic DNA is copied during cell divisions. These strange parasites are discussed in more detail in Chapter 6. Since most of this noncoding DNA is never translated into protein, mutations in this DNA often have no significant effect whatsoever on the organisms involved, and since this general category far exceeds the coding DNA, this is where much neutral evolution occurs.

On a cautionary note, our knowledge of this noncoding DNA is still quite recent, and it is still uncertain what fraction of this DNA is without effect on the final phenotype of the organism. In fact, several recent findings suggest that at least some of this noncoding DNA does have important regulatory functions, so mutations in at least some of this noncoding DNA may affect the fitness of the organism. It will undoubtedly take a few more years of work before a confident estimate of the relative percentages of coding/regulatory versus "junk" DNA is available. For now, it still appears that there is significant noncoding DNA in the genome within which most mutations would likely be neutral.

There are other classes of noncoding DNA that likewise can be mutated without affecting organismal traits. Pseudogenes are numerous in many eukaryote genomes. These are usually extra copies of useful genes (they came about by gene duplication chromosomal mutations) that have since mutated into useless and inactive stretches of DNA. The organism needs only one copy of the essential gene, so extra copies can be mutated to the point of uselessness, yet still remain a part of the genome. Humans now carry 49 useless pseudogenes of the cytochrome C gene alone. About 20,000 pseudogenes have been identified in the human genome in total—almost as many as the coding genes (Fairbanks, 2007). Most organisms also carry a few vestigial pseudogenes. These are often single gene copies that mutated to nonfunctioning states after their function was no longer essential to the organism. In this situation also, the unnecessary gene will simply accumulate more and more neutral mutations with no effect on the organism's fitness. It is thought that this process explains the loss of eyes in some cave-dwelling species of fish and salamanders. In a cave population of fish, if a mutation shuts down a gene necessary for constructing functional eyes, the mutation would most likely be neutral—even slightly beneficial since it might prevent a waste of calories and nutrients that would have been used in constructing useless eyes—as cave environments are typically completely dark. Since a great deal of evolutionary change is due to trait loss, many vestigial pseudogenes are present in the genomes of most organisms—especially so in the more complex eukaryotes, and continued mutations in such genes contribute to further neutral evolution of the species.

SOME CHROMOSOMAL MUTATIONS CAN BE NEUTRAL

Beyond mutations in single genes or in classes of noncoding DNA, chromosomal mutations also occur from time to time. These involve various deletions, translocations, duplications, or inversions of small to sizable segments of chromosomes (Chapter 5). Some of these are clearly likely to be fatal or detrimental, as when a large segment of a chromosome containing essential genes is deleted and not passed on to offspring. Some chromosomal mutations, however, result in genes being duplicated, which will often be neutral in effect. Some translocations of a chromosome segment from one chromosome to another can likewise be neutral as long as all needed genes are still present somewhere in the genome. This raises again the important point that genomic evolution is not just limited to the genes themselves, but also includes the relative order and arrangement of genes and other recognizable units of DNA within and between chromosomes—sometimes referred to as the "genome architecture". Just like alleles, some chromosomal rearrangements can also become fixed in one species, but not in several related species; this genomic feature then becomes an important and measurable evolved genomic trait, with potential usefulness in phylogenetic comparisons and cladistics analysis.

These five categories cover some of the major possibilities that can lead to neutral evolution. Actually, any mechanism other than natural selection (that leads to the elimination of unfit variation) or the known mechanisms that typically lead to decreased fitness (like some transposable elements, B chromosomes) can be said to be factors in neutral evolution. Genetic drift, the topic of Chapter 9, is one example of genomic change that can be neutral because it mainly affects neutral or nearly neutral alleles. Another example is where a newly inserted bit of neutral DNA is driven to fixation in a species because it is positioned very near (linked to) an allele that is increasing in the population due to its positive effect on fitness. In this special case, even natural selection can give rise to neutral genomic change by "dragging" the linked neutral DNA to a higher frequency in the population—possibly even to fixation.

There may still be other undiscovered or unrecognized categories of neutral evolution hiding in the genomes of the millions of species inhabiting our planet. Based on what has been learned in recent decades, neutral evolution is unquestionably a huge contributor to the evolution of genomes, yet one that, unlike natural selection, cannot create adaptations in organisms except by the most indirect of routes. The best known example of this is when duplicated genes (initially a neutral mutation) become mutated in such a way as to code for a different kind of protein that does have functionality for the organism involved—as was discussed in Chapter 5.

Chapter 9

Genetic Drift

In a small population alleles may be lost simply through errors of sampling (stochastic processes); this is known as genetic drift.

Ernst Mayr

Genetic drift has been recognized as a force in evolution for almost a century. In older textbooks, it was often the only evolutionary mechanism to be discussed in addition to natural selection. Genetic drift can be briefly defined as the random fluctuations of allele frequencies from generation to generation that are not due to natural selection. It is also know as "random drift in allele frequencies". Because of this random genetic drift, even a new or rare allele might come to dominate a population over several generations, especially so in small populations. The eventual end point of genetic drift is that some alleles will eventually become fixed in a population, with alternate alleles being eliminated completely.

Of course some alleles are selected against by natural selection, and these usually tend to be present in relatively low frequencies due to that negative selection pressure. Many alleles, however, are neutral or near-neutral in their effects on the phenotype. For instance, is there any adaptive advantage to those two earlobe shapes mentioned in Chapter 8—attached or unattached? Almost certainly there is not. Genetic drift is considered to be a more noticeable force with neutral alleles, since selection acting against harmful alleles would tend to exert a more powerful influence on allele frequencies than random drift. However,

Evolution. http://dx.doi.org/10.1016/B978-0-12-800348-0.00009-2
Copyright © 2014 Elsevier Inc. All rights reserved.

many alleles that are "selected for" are only slightly advantageous compared to alternate alleles, and in this case genetic drift can still work as an additional factor affecting those allele frequencies from generation to generation.

The effect of genetic drift can be especially potent in what are called founder populations. One proposed example of founder effect genetic drift concerns the small founder population of humans that crossed the Bering land bridge some 15,000 years ago to subsequently give rise to the Native American populations in North and South America. Evidence indicates that this small band of wanders by chance carried essentially no B alleles of the ABO blood group. In fact, even the A allele was missing or extremely rare, so that until fairly recently most American Indian populations expressed almost exclusively type O blood. Eskimos and a few other native groups in northern North America do contain some A alleles, which may be due to a second crossing of the land bridge by a different band of nomadic people in more recent times.

A more recent example of a founder event occurred around 1945. At that time African cattle egrets showed up in Florida. Somehow a few individuals had become lost or caught in a storm (we will never know the exact details) and ended up surviving the Atlantic crossing to Florida. The number that made this crossing was small and certainly far less than 1% of the parent egret population back in Africa.

Being such a tiny part of the original African population, these few individuals certainly did not contain all the various alleles present in the larger "parent population". Some alleles that were rare in the parent population were undoubtedly absent in this small founder population. Conversely, this founder population may have contained at least a few other alleles that were rare in the African population, but are now very common in the North American population simply because they happened to be more common in that small founder population. This small band of immigrants has grown dramatically since 1945 to cover much of North America, and they now number in the millions in the Americas. The establishing of a new population with different allele frequencies compared to the larger parent population is an especially rapid and dramatic example of the "sampling effect" of genetic drift. To my knowledge, no genomic comparisons have been done between the parent population in Africa and the relatively new population in North America, but a comparison would be expected to turn up some specific examples of genetic drift due to the founder effect.

One example of a bottleneck effect occurred with the American bison or buffalo. Buffalos existed in the millions until they were hunted to near extinction by non-native hunters in the 1800s. By the late 1800s there were only 2,000–3,000 buffalos left. Through several recovery efforts since that time, their numbers today are back up to over half a million. Certainly some alleles were lost completely, and many allele frequencies were shifted dramatically in this near-extinction bottleneck.

I will now use the well-worn analogy of a bowl of colored marbles to illustrate genetic drift. If you had a large bowl filled with 90% white marbles and only 10% red marbles (to represent two alternate alleles of one gene), if asked to withdraw only four marbles as a founder population (imagine marbles can reproduce), you would not be surprised to withdraw four white marbles and no red marbles. In this case, the new founder population would have lost the red "allele" altogether and the white allele would be instantly fixed in that founder population. In far rarer instances, you might draw two red marbles and two white marbles. In this instance, the founder population would have moved from 10% red to 50% red alleles, and over a few generations random genetic drift could potentially even push the red allele higher. In either of these possibilities, a significant change in allele frequencies would have occurred in the new and perhaps isolated population. It has been proposed that a mix of random genetic drift plus new selective pressures (on a variety of genes and alleles) in a

relocated founder population can potentially lead quickly
to the start of subspecies formation, and eventually speciation given enough time.

Back to that bowl of marbles, remember that the white
and red alleles represent two alleles in only one gene, and
there would of course be thousands of other genes—many
of which affect the survival and reproduction of the next
generation. Even without isolation of a founder population, only a few of the individuals produced in any generation survive to give rise to the next generation. Even
among complex animals like African cheetahs, only about
one in seven offspring typically survive to produce the next
generation. In some insects it is only one or two out of
100 or 1,000 eggs that produce the next generation. In our
marble example, if significant numbers of individuals carrying "the red allele" die young for reasons unrelated to
that allele, the frequency of the red allele will go down.
Alternatively, if by chance several individuals carrying the
red allele survive to produce the next generation, the frequency of that allele may go up—again for reasons possibly unrelated to that allele's effects (if any). This again is
genetic drift. It is an ever-present factor in all populations,
and given hundreds to thousands of generations in a species where only a small percentage of the young survive to
reproductive adulthood, its effects can be significant even
in large populations.

One last example and point: if again there are 10% red
alleles and 90% white alleles in generation one, the next
generation (generation two) could easily be made up of
12% red and 88% white alleles due to genetic drift. Hopefully you would see that the chance of the next generation jumping to 14% red and 86% white is now greater
than would be the case if starting again from a 10/90 split.
In short, each time an allele frequency climbs higher, its
chances of climbing even higher due to drift are enhanced
slightly due to its now greater frequency (a ratcheting
effect). Alternatively, if the frequency of an allele declines,
its chances of declining even further are enhanced. Over a

great many generations, such chance processes will probably lead to the loss of one allele and the fixation of the alternate. We now know that many such fixations are of neutral base substitutions in the genes, and fortunately such substitutions can today be identified and then used as clues in working out phylogenies—the course of evolution within a clade.

Suppose there are seven species of ferns in the same genus (species A–G), and we find that in a section of a homologous gene, five of the species have the codons reading AAG CGA TCA, while species B and E have the reading AAG CGT TCA. The neutral substitution of a T for and A in the second codon (the codon still codes for alanine) is most likely explained as a homologous condition between these two species. Though we would never base a final phylogenetic decision on a single bit of evidence like this, if this was all the information we had to go on, we would make the logical assumption that the two species B and E sharing this substitution are sister species and are therefore more closely related to each other than either is related to the other five, which lack this substitution. Especially since this is a completely neutral change, its fixation in the ancestor of species B and E is most likely due to past genetic drift rather than to natural selection.

Today many hundreds of genes are being read and compared between species and groups of species in higher taxa to identify phylogenetic relationships and thus work out ancestries, and one of the important tools in these studies involves the use of neutral base substitutions that have become fixed through genetic drift in some species, genera, families, orders, or even higher taxa. In addition to coding alleles, mutations in introns, pseudogenes, and retroelements, and even some chromosomal mutations can also undergo genetic drift over time, and they too can and have become fixed in some ancestral lines—allowing us even more sources of evidence for use in working out phylogenetic relationships.

A final point that perhaps needs to be covered here is that the idea of genetic drift seems to be in opposition to the Hardy–Weinberg law, which states that if alleles are neutral, and mating is random with respect to those alleles, then the frequency of the alternate alleles should remain relatively constant and resistant to change over the generations (actually there are several other qualifiers of this law, such as large population size and lack of movement of individuals in or out of the population). Genetic drift seems to suggest the opposite outcome—that allele frequencies will randomly drift over time. The Hardy–Weinberg law of course applies to an idealized state that almost never holds true in nature—and even under those ideal conditions the law states only that the alleles will *probably* remain in the same relative frequencies from generation to generation. It does not state that they definitely will.

This is similar to saying that probability predicts we should get 50 heads and 50 tails out of 100 coin tosses. In practice one would probably be surprised if 100 coin tosses actually gave those expected results. Fifty-five heads and 45 tails would be unsurprising, as would allele frequency changes of this magnitude, especially in smaller populations or in populations of a larger size where few of the produced offspring survive to reproductive age—as is the case for a great many of the Earth's species. Over a great many generations these random changes in frequencies due to pure chance will eventually remove one of the alleles and fix the other. Again, it will also fix some of the many neutral mutations in noncoding DNA or in chromosome rearrangements in some populations and lines as opposed to others. This is why all the apes (including humans) have cytochrome C that is one amino acid different from that of most monkeys. This is a neutral amino substitution that most likely originated in the ape line of descent after "we apes" split from the monkey line. It was at first a rare allele coding for an equally functional but different molecule. It then eventually became fixed in the ape line over time and so has been inherited by all modern apes and humans.

Chapter 10

Environment

Natural selection is, first of all, a theory about adaptation to changing local environments, not a statement about "improvement" or "progress" in any global sense.

Stephen J. Gould

The same genotype may produce quite different phenotypes in different environments.

Ernst Mayr

The environment determines whether novel phenotypes become innovations, or whether they perish.

Andreas Wagner

Evolution in the long run depends greatly on environmental change. This can happen through actual environmental change (which we know has occurred extensively over much of the Earth over geologic time), or it can occur through the displacement of organisms into new environments—as when seeds, insects, or birds arrive on isolated oceanic islands or disjunct continents. Without environmental change, natural selection would eventually "perfect" (as much as possible given the available genetic variation) each species to be well-adapted to its respective environment and lifestyle. Further adaptive evolution would then have to await the appearance of even more superior genes/alleles through mutation. In short, evolution would slow to a crawl, resulting in near evolutionary stasis. Much of the stasis referenced by punctuated equilibrium theory is likely due to just such situations.

Evolution. http://dx.doi.org/10.1016/B978-0-12-800348-0.00010-9
Copyright © 2014 Elsevier Inc. All rights reserved.

Some species seem to have attained evolutionary stasis by virtue of continually living in environments that have changed very little for vast spans of geologic time. The chambered nautilus of today is virtually identical (as far as we can tell from their fossils) to those of nearly 400,000,000 years ago—which is almost two-thirds of the time for which we have a good fossil record of animals. Its home environment being the relatively stable deep sea most likely explains the nautilus's early attainment of near-adaptive perfection and continued long-term stasis.

Horseshoe crabs have a similar record of relative morphological stasis over almost the same period of time. They inhabit the shallower marine habitat of the continental shelves, but again their fossil record indicates that horseshoe crabs long ago became well-adapted to existing and thriving on continental shelves, which have apparently remained relatively stable environments—at least in terms of those factors important to horseshoe crabs. Of course, both the nautilus and the horseshoe crab have most likely continued to evolve somewhat in their physiology, biochemistry, and behavior, but those changes cannot be elucidated from the fossil record, which records only their morphology. A great many other organisms appear to illustrate long-term stasis, including scorpions, roaches, ginkgo trees, and cycads. In some of these cases, like the scorpions and roaches, cladistic evolution certainly did occur through numerous speciations in more recent times, but without major morphological changes to the fundamental body plan of these animals.

Though we have essentially no fossil record for protists like the many forms of amoeba, it is a pretty good bet that they too, being single-celled organisms with a simple flexible cell membrane, are organisms that have changed little over even greater spans of geologic time. Watery environments filled with plenty of prokaryotes have certainly been present on the planet for most of the past two billion years, and it is this type of environment in which most free-living amoeba thrive. For prokaryotes, we do have some very ancient impressions of filamentous cyanobacteria that look amazingly similar to modern forms. These cases of exceedingly long periods of stasis do occur in certain groups, whereas many other groups have undergone rapid and drastic evolutionary change over much shorter periods of geologic time. For example, whales evolved from far smaller land animals in less than 50,000,000 years. Birds evolved from theropod dinosaurs and have continued to diversify and change within the last 170,000,000 years. Our human "family" (now considered a taxonomic "tribe" of more than a dozen known species) emerged and diversified into several species within only the past 6,000,000 years, though only *Homo sapiens* has survived this recent flourish of speciation.

In those examples of significant stasis, it is probable that the major features of the environment of these groups have been relatively stable and continually present—allowing these groups to survive over vast stretches of geologic time without the need for major evolutionary changes. In the contrasting examples of more rapid change in whales, birds, and humans, the environmental changes that allowed or encouraged evolutionary changes may never be fully known, but they may have included the sudden appearance of vacancies in "environmental space". For instance, some argue that whales may have evolved in part because the marine environment had been recently vacated of the large monosaurs, plesiosaurs, ichthyosaurs, and other large reptilian vertebrate predators that suffered extinction along with the dinosaurs some 65,000,000 years ago. This mass extinction created the opportunity for some vertebrate group to enter these vacated ecological niches and "refill" them yet again. More commonly mentioned is the related example of the rapid evolution and diversification of land mammals after the majority of the dinosaurs suffered extinction in the terrestrial environments, leaving again a lot of potential ecological opportunity—or unfilled ecological niches in the parlance of ecologists.

Early birds were likely entering ecological space that was formerly unfilled by flying species of that size—at least in some inland areas where small pterosaurs were not common. The split of humans from ancestral apes is often argued to be due in part to climatic changes in central Africa, while later human evolution in Europe and Asia was affected by the several more recent ice ages. In these and many other known cases of evolutionary change and adaptation, environment most likely played a major role.

Environmental change often opens up new opportunities for species to enter and adapt to those new environments. One singular example involves the rise in the level of Lake Malawi in the eastern African Rift Valley region. The water level of this large lake has risen significantly in only the last few hundred years. Along the central eastern shore there are now a few clustered rocky islands that only 200–300 years ago were dry land along the shoreline. Amazingly, more than 20 species of cichlids (freshwater fish) are now endemic to the waters surrounding those islands (Schilthuizen, 2001). It seems that this new environment was quickly seized by one or more "ancestral" cichlids that speciated repeatedly and rapidly in adapting to the several newly available microhabitats in that area. This example is repeated in Chapter 13 as one of the dramatic examples of sympatric speciation.

Even more drastic environmental change occurs when individuals of surface-inhabiting species are swept or fall into deep and large cave systems. Around the world we have records of fish, salamanders, arthropods, and worms that have been able to adapt to the dramatically different environments of these caves—all of which have some things in common. Since most caves are completely dark, well-adapted cave species have, at least to some extent, lost their eyes and their skin pigments, neither of which serve any role in a world of perpetual darkness. These animals will also often evolve a small and spindly body form as an adaptation to the smaller amount of food available in a cave environment (no photosynthesis to provide a large food base). Also, any former daily or yearly cycles of activity will eventually disappear since the cave environment does not vary significantly on a daily or yearly basis.

Ecologists typically describe the environment as having three major constituents: other members of one's own species, members of other species, and those abiotic or nonliving factors in the environment like temperature, wind, salinity, and others. Since each of these three components is important in countless ways to most species, evolution shapes species to be able to deal with the many various and likely interactions within each of these components. Likewise, when factors in any of these components change, some degree of directional selection is likely to occur (or extinction in some cases). In terms of interactions within the species (intraspecific interactions), competition and sexual selection (the latter only in sexually reproducing species) are two of the more important factors. Interspecific factors include an exceedingly larger range of possible interactions, including competition, food web factors, symbiosis in one form or another, habitat changes, and many more. All biotic environmental factors, whether intraspecific or interspecific, are ever-present and potentially rapidly changing as opposed to the usually more slowly changing abiotic environmental factors. Rarely, abiotic factors do change rapidly, as when climate is disrupted by large-scale volcanism, but more often they change slowly as when continents drift slowly to different latitudes. Some abiotic factors do change somewhat drastically, but in regular and repeating cycles that species can adapt to handle, with seasonal changes in temperature and light levels being one major example. Known examples and details of change under the three fundamental environmental categories could easily fill several volumes.

Even when evolutionary change is apparent, typically only a minority of the traits of a species are changing, with most remaining the same. If at any one time too large a number of the traits of a species were not well-adapted to the available environment, it is unlikely the species could survive, and extinction would be a more likely outcome. To put it more technically, the great majority of traits at any one time are under stabilizing selection, while a few may be undergoing directional selection toward a more-adapted state. In the famed peppered moth observations of Henry B.D. Ketterwell in England, only the genes affecting wing color were observed to be evolving from the peppered form to the black or melanistic form in forests where air pollution had killed the lichens against which the peppered form was so well camouflaged (Rudge, 2005). As far as we know, other aspects of the moths were still well adapted during this time of directional selection affecting the wing and body color. The environmental change in this classic example was brought about by the air pollution around industrial centers in England, which killed the light-colored lichens that grew profusely on the trunks of forest trees. This change made the peppered moths stand out clearly against the now-exposed dark bark of the trees, and the moths were no longer camouflaged from their bird predators. The occasionally appearing melanistic forms of the peppered moth were then favored since they were less noticeable to the birds than the previously well-adapted lighter-winged individuals. Directional selection had set in at this point to eliminate the lighter-winged peppered moths and leave the uniformly dark individuals.

The role of environment in evolution takes on a significantly different aspect in the world of microbes. Because of the various possible mechanisms that allow for horizontal gene transfer (Chapter 7), genomic evolution is influenced much more directly and rapidly by the environment than is typically the case for eukaryotes—and especially multicellular eukaryotes. In the case of conjugation among prokaryotes of different species, or transduction involving viruses, or transformation involving environmental DNA (which had been freed from other organisms in the environment), genomic evolution will depend in part on what other species of bacteria, archaea, protists, and viruses are present in the same environment, as they will be the most common sources of potential DNA gains through horizontal gene transfer.

The simple fact that our planet's environments are so diverse and numerous explains much of the richness of biodiversity. In the marine realm alone there are vastly different and distinct environments. The differences between a tropical coral reef environment and the deep sea floor, or a hydrothermal vent area, are really huge. Not even the microbes inhabiting these three environmental have much in common, and certainly the larger forms show extreme differences in morphology, physiology, pressure and chemical tolerances, etc. Environmental diversity is one of the major reasons why the first life form ever diversified at all—starting the beginnings of the branching tree of life.

Environment is simply key to evolutionary change, something that Darwin recognized more so than most of his contemporaries. Evolution is not innately driven as Lamarck and many others had believed; it is rather environmental pressure and change that bring on the evolutionary adaptations in species through the process of natural selection, and even nonadaptive changes due to genetic drift can often be indirectly tied to environmental changes that, for instance, result in decreased population sizes and bottleneck effects. Simply put, evolution does not occur in a vacuum, it occurs in a complex and multifaceted environment, one that will likely change in numerous ways given sufficient time.

Chapter 11

Development

Evolution is not a genetically controlled distortion of one adult form into another; it is a genetically controlled alteration in a developmental program.

Richard Dawkins

Most probably the relevant steps towards the evolution of higher forms, humans among them, have arisen in regulatory and assembling processes that decide when, where, and in which combination the already existing genetics blocks operate.

Antonio Fontdevila

As evolutionary theory has aged and matured, more and more factors have been recognized as integral to evolution—with each being incorporated in its measure into the expanding evolutionary paradigm. First it was the obvious comparative morphology of organisms, then classical genetics, then population genetics and genetic drift, then neutral evolution, etc. The area of developmental biology—especially comparative developmental biology, was recognized early on as a source of evidence for evolutionary theory due to the great similarity of embryos within some classes and phyla—especially those of the vertebrates. This similarity was clearly suggestive of descent with modification. Additionally, the occurrence of ancestral structures in the embryos of animals that lacked those same structures as adults was instructive in regard to ancestry and phylogeny. Examples would include hind limb buds in the embryos of whales, significant tails in the embryos of apes and humans, and the gill slit folds

Evolution. http://dx.doi.org/10.1016/B978-0-12-800348-0.00011-0
Copyright © 2014 Elsevier Inc. All rights reserved.

in the embryos of all terrestrial vertebrates. But only in recent decades have we begun to understand that developmental processes and their associated genetics represent a substantial wellspring of potential evolutionary change and diversification—and one that fortunately also incorporates a record of many past evolutionary changes.

Developmental processes have obviously contributed little to nothing in terms of evolutionary diversification for the prokaryotes and the majority of protists. But for those more complex organisms that develop multicellular bodies—often from embryos as in the plants and animals, a great many developmental possibilities exist for modifications in the form and function of the adult organism. This is because a significant number of genes and regulatory DNA sequences take part in the symphony of activity that constitutes development. While some genes code for the actual building block molecules of cells, a great many others control the activity, timing, and rate of the many formative developmental events that occur as the initial zygote (fertilized egg) develops into the fully formed organism. Alter just one crucial gene responsible for controlling the timing of an essential developmental process, and the results can be dramatic. If you are familiar with the old saying "as the twig is bent—so the tree is inclined" you may by analogy be able to glimpse the importance of regulatory gene activity in controlling organismal development and its outcome in terms of adult morphology and physiology.

The importance of modern developmental biology to evolutionary biology stems from a number of significant discoveries and technologies of recent years. Specifically, the ability to sequence whole genomes of organisms, the ability to discover which regions of the genome are coding or regulatory, and the ability to compare these DNA sequences across higher-level taxa have vastly increased our understanding of how evolutionary changes can emerge from genomic and developmental processes. One of the most surprising and enlightening discoveries was the widespread commonality (homology) of the genetics behind the development and form of very different organisms. Before genes could be identified and sequenced, it seemed sensible that animals as different as a fly and a dog were different because they were the result of drastically different genomes—different genes resulting in different biomolecules, cells, and thus animals. This is still true to some degree, but not nearly to the extent formerly believed. Instead we find very similar genes in organisms as divergent as flies and dogs controlling the development and formation of the animal and its components in generally similar ways—but "tweaked" enough to result in distinctly different outcomes. These homologous genes responsible for common themes of development across large clades of species (like the entire animal kingdom) have been dubbed "the genetic toolkit" by many of the workers in this discipline—same toolkit, but used differentially to bring about different results across species.

These different results would include items like a longer femur in one species—a shorter femur in another; a long tail in one species—a vestigial tail in another; dark coloration in one species—lighter or mottled coloration in another; legs on four body segments—legs on only three body segments. All these variations and more are possible from the same genetic toolkit. Many of these toolkit genes (known as homeotic genes) produce proteins known as "transcription factors". Transcription factor proteins act as switches that bind to regulatory regions of DNA near the start of other genes. Then, due to this binding, the transcription factors either stimulate or inhibit the transcription of mRNA from the affected gene. In plants, one group of transcription factor genes is known as the MADS-box gene family. This gene family contains two to five transcription factor genes, depending on the plant group. In most of the flowering plants this gene family determines flower structure, with differential activity of the five MADS genes causing some flower "whorls" to become sepals, some petals, some stamen, some carpels, with the fifth gene stimulating formation of the ovules within the carpels (Bergstrom and Dugatkin, 2012). Most plants use this set of toolkit genes, though some of the more ancestral nonflowering plant groups get by with only two or three MADS genes performing different functions.

Though perhaps not a perfect analogy, consider the switches that control the connections of train rails coming out of train switching yards. A switch in one setting sends a boxcar to Chicago, the alternate switch setting sends it to Detroit, and another switch farther down the "Detroit line" line might switch the whole train to Buffalo. So given the same rail cars, the same rails, and similar switches, you get different destinations depending on the relative settings within the switching system. In a somewhat similar but more dramatic way, different but related species have come about in part through differential programming of the many transcription factor genes within their respective genomes. For complex organisms with multiple switching genes, the potential for such developmental divergence is great. The larger the number of genetic switches, the more variety is possible in phenotypes, and most of the well-known groups of genetic switches have come about through gene duplications such as the ones that gave rise to the homologous MADS genes, and the even larger number of homologous HOX genes in animals.

The most often used example of developmental switching genes is the family of HOX genes found in most animals. Because of past bouts of gene duplications, vertebrates have what are recognized as four sets of duplicated HOX genes, and within each set there have been some further duplications—as well as a few gene losses. The HOX gene switches act in complicated pathways that vary in different body regions to control the specific development of those regions (arranged from anterior to posterior). The result of their differential actions along the anterior/posterior gradient is the development of the many phenotypic traits specifically associated with certain regions of the head, the neck, the trunk, and the tail of the animal. Such traits would include the correct morphology of the vertebrae in that body region, the correct organs in the correct location, the correct muscles in a specific body region, modifications of the nervous system appropriate to that body region, etc. The finding that HOX genes go back even into sponges is an amazing discovery, so they have seemingly been with the animal kingdom since its inception, yet evolving in their own right along each animal line of descent.

A powerful example of a relatively small developmental tweak with profound results can be found in the Mexican salamander *Ambystoma mexicanum*: common name—the axolotl. It turns out that this species undoubtedly derived from the tiger salamander *Ambystoma tigrinum* as the result of a single developmental regulatory change that stifled the normal burst of thyroid hormone (thyroxine) that would have caused the animal to metamorphose into an adult tiger salamander (Bergstrom and Dugatkin, 2012). The axolotl does become reproductively mature, but it retains the larval morphology suited for an aquatic life, including large functional gills, a laterally flattened tail for swimming, and larval coloration. In a sense, the axolotl is now a reproductively mature larva due to the "downregulation" of a gene responsible for the thyroxine burst. This change is believed to have originated in an ancestral population of salamanders living in an environment where there was more danger from land predators than from aquatic predators. In short, it benefited the species to remain water-adapted rather than metamorphose into an animal more suited for terrestrial living. So this dramatic evolutionary change was the result of meager mutational change that resulted in a dramatic (and likely rapid) change in the line leading to the modern axolotl.

A great many other developmental changes have given rise to evolutionary novelty by way of the loss or reduction of a former phenotype. This could result from the deletion of a nonessential gene, but more commonly results from mutations that cause a downregulation of a former gene-to-phenotype pathway. In the case of vestigial structures like the much-reduced wings of a kiwi bird, the downregulation was extreme, though a tiny adult wing is still produced. In other cases the switch is blocked or completely turned off by mutations and no end phenotype is produced. There is evidence that this is the case with teeth in birds. Ancestral birds had teeth similarly to the dinosaurs they had recently evolved from. Later, birds lost their teeth completely, yet they still have "teeth genes" and/or upregulating switches that have now mutated into nonfunctional stretches of DNA.

The development of eyes is typically disrupted in the evolution of cave-inhabiting fish, salamanders, crayfish, etc. The once upregulating homeotic genes, or their target genes that cause eye development, are gradually mutated into dysfunctional stretches of DNA—without selective cost in the darkness of the cave. Several parasite groups serve as good examples of animals that have likewise lost eyes (none needed in the darkness of an intestine, liver, or lung). Both tapeworms and flukes evolved from free-living flatworms that like their modern counterparts usually do have a pair of simple eyes. Locomotor mechanisms are also often cut back or eliminated in some parasitic groups. Free-living copepods have appendages used in swimming, but many of the adult parasitic copepods do not develop these appendages since they spend their adulthood permanently attached to their vertebrate host—usually a fish.

There is still undoubtedly much to be learned about the interface of developmental biology and evolution, but clearly we have learned enough to know that a great deal of evolved biodiversity is the result of the duplication and eventual differential evolution of various homeotic genes (and their target genes) resulting in expanding (and sometimes shrinking) developmental networks of great complexity—both within and across species. In this chapter, only one mention of a life cycle developmental change was given—that of the tiger salamander to axolotl. This was a simplification of a life cycle to eliminate the typical terrestrial adult stage, but vast numbers of animals have numerous developmental stages in their life cycles, especially so in some parasitic groups. The evolution within a parasitic species of a distinct developmental stage in three or more different phases of its life cycle is clearly another major change involving an interplay of development and evolution, and one that is only starting to be addressed and elucidated.

Chapter 12

Symbiosis

We are here thanks to successive endosymbioses.

Alexandre Meinesz

Some obligate bacterial parasites and mutualists have genomes so small and lifestyles so dependent on a host that they are at the interface between organism and organelles.

John N. Thompson

Today, entomologists think that virtually every insect known to science is likely to contain symbiotic microbes, usually including some in their guts.

Tom Wakeford

Only in recent years has the important role of symbiosis in evolution become apparent to a significant number of biologists. Of course biologists have always known of the countless examples of species-to-species mutualisms, commensalisms, and parasitisms—with almost all living species involved in one or more of these types of relationships, and with many of these relationships showing obvious signs of coevolution between the species involved. Dr Lynn Margulis has argued that symbiosis is a major feature of evolution and has even claimed that "most evolutionary novelty arose, and still arises, directly from symbiosis". This depends on what one means by novelty. Surely the speciation of the many current whale species from a common ancestor is not obviously due to any unique symbiotic relationships that drove each of

Evolution. http://dx.doi.org/10.1016/B978-0-12-800348-0.00012-2
Copyright © 2014 Elsevier Inc. All rights reserved.

those many speciation events, and the same holds true for most other cases of speciation in the "lower" taxa.

Symbiosis does, however, seem to have been involved in at least the origins of several diverse, unique, and successful groups such as eukaryotes—where at least mitochondria and chloroplasts are known to be derived from once free-living prokaryotes. The Serial Endosymbiont Theory of eukaryotic cell origins is now well accepted and is one of the most important and dramatic examples of the role of symbiosis in the evolution of life on Earth. According to this theory, the eukaryotic cell evolved in part as the result of endosymbiotic unions of previously free-living prokaryotic cell types with other existing cells. Most evidence now supports the "host" cell in this development as having been from a line more closely related to the archaea than to the bacteria. This host cell engulfed or in some way incorporated internally a purple alpha-proteobacteria that eventually evolved into mitochondria within the host cell line (Shapiro, 2011). This endosymbiotic event is believed to have occurred around two billion years ago, though surely it took a great number of generations to work out the mutualistic arrangement to the point of being able to call the endosymbionts true organelles—mitochondria in this case. Some suggest that originally this was a parasitic relationship that gradually evolved into mutualism, but we will really never know with any certainty the exact pathway that this relationship traversed.

In one or more later eukaryotic lines, a cyanobacterium (a photosynthetic prokaryote) was likewise engulfed by a eukaryotic cell already containing mitochondria—giving rise eventually to the first algae, with the host cell utilizing some of the carbohydrates manufactured by the cyanobacterial endosymbiont. These cyanobacterial endosymbionts eventually evolved into chloroplasts (Shapiro, 2011). This event is believed to have taken place as recently as 1.2 billion years ago, and to have given rise eventually to the red algae and the green algae.

Again, both of these two known symbiotic events most likely required a great many generations to work out the relationship to the level of refinement we see in modern eukaryotes. For the endosymbiont to evolve into a true organelle, it must be both retained by the host cell and reproduce "within" the host line of descent—as opposed to only being incorporated into individual host organisms, with each new generation of host organisms having to recapture the endosymbionts from a free-living population of organisms. Only when the endosymbiont can reproduce "vertically" from host parent to offspring can it really evolve into a fully dependent organelle of the host cell. This is a major refinement in the relationship that then leads to other adjustments in the relationship. One major adjustment typically involves genome reduction in the endosymbiont, with some now-unnecessary genes being lost (unnecessary because the host cell genes duplicate some of the functions of those in the new endosymbiont). Another common outcome is for some of the essential genes from the endosymbiont to be transferred over time into the host cell nucleus. This latter process can eventually result in only a few genes remaining in the endosymbiont (the mitochondria or chloroplast). Human mitochondria for example now retain only 13 protein-coding genes from their original store (Avise, 2010).

Around 18% of the nuclear genes in the plant genus *Arabidopsis* are of cyanobacterial origin—almost certainly derived from their chloroplasts over evolutionary time, but because chloroplasts arose from a more recent endosymbiotic union, they often contain a good deal more of their original genome than is the case with mitochondria. Some algal chloroplasts still contain from 100 to 300 functional genes, but this is still far fewer than the 500 to 1,000 typical of free-living cyanobacteria.

Lynn Margulis also believed that cilia and flagella (undilipodia) were the result of an even earlier union of "the Achaean" and a spirochete bacterium, but this hypothesis is not as well supported and remains unconvincing to most biologists. All such events that give rise to a new organelle in the cell line are called primary endosymbiotic events.

The origins of photosynthetic euglenophytes (includes the well-known genus *Euglena*), diatoms, golden algae, and brown algae happened later in time by secondary endosymbiosis events, wherein a photosynthetic protist (a red or green algae) was engulfed by or incorporated into a heterotrophic protist. In this event, both the host cell and the new endosymbiont were eukaryotic cells (Katz and Bhattacharya, 2006). The "engulfer cell" became the dominant cell line, with the engulfed endosymbiont slowly being reduced to essentially its photosynthetic chloroplasts, which are now characterized by three outer membranes (rather than two as in the green algae and red algae that resulted from primary endosymbiosis). The third membrane is in most cases believed to be the remains of the outer cell membrane of the engulfed cell, which has "shrunk" around the chloroplasts.

In some examples of secondary endosymbiosis, even the nucleus of the engulfed protist is still present as a nucleomorph—essentially a small vestigial nucleus of the endosymbiont eukaryote that continues to be copied and passed on in the lineage along with the functional chloroplasts (Katz and Bhattacharya, 2006). The nucleomorph exists in a small membrane-bound pool of cytoplasm that also encloses the chloroplasts of the endosymbiont. In other lines, the nucleus of the endosymbiont has been lost and only the chloroplasts remain within the extra membrane. Because of all these endosymbiotic unions, some modern eukaryotic cells can be described as genetic chimeras containing genes from:

1. an archaea-like prokaryote (original host cell)
2. a purple alpha-proteobacterium (mitochondria)
3. a cyanobacterium (in the case of red and green algae and plants)
4. a captured eukaryote (in the case of brown algae and diatoms).

If this story so far is not already confusing enough, a few of the photosynthetic dinoflagellates (not all) are believed to be the result of a tertiary endosymbiosis, wherein a heterotrophic dinoflagellate engulfed a photosynthetic eukaryote of secondary endosymbiotic origin and began a relationship of symbiosis with this new endosymbiont.

In short, all eukaryotes (domain Eukaryota—many millions of species) are the result of primary, secondary, or even tertiary endosymbiotic events. Surely these ancient and extreme cases of symbiosis qualify as major contributors to evolution on planet Earth.

In recent years we have gained evidence of other endosymbiotic unions that apparently occurred more recently and are thus restricted to smaller groups of organisms. One good example is found in some of the recently discovered species of clams that live around some of the deep-sea hydrothermal vents that occur where two crustal plates are forming and separating as magma pushes up and solidifies between them. It has been discovered that the majority of the nourishment for these clams comes from chemosynthetic bacteria that have been incorporated into the cells of the clams—especially into the cells of their gills (van Dover, 2000). These chemosynthetic bacteria utilize chemicals like hydrogen sulfide (present in the water spewing from these vents) to create energy through oxidation reactions. The energy released from oxidation of the hydrogen sulfide substitutes for sunlight in this dark world at the ocean floor, allowing these bacteria to manufacture hydrocarbon fuel molecules like carbohydrates, some of which are passed on directly to the host clam cells.

Importantly, these bacteria are passed vertically to the next generation of clams in their eggs. Clams house their eggs within hollow cavities within their gills before releasing them into the seawater. The bacterial symbionts enter into the developing eggs before their release so the next generation will automatically contain them. These particular bacteria are only known from clam tissues—they do not exist freely in the seawater around these vents. It has also been discovered that these endosymbiont bacteria have much-reduced genomes compared to other types of free-living chemosynthetic bacteria in the hydrothermal vent environment. All evidence points to the now-familiar process whereby these bacteria were somehow taken in and moved down the path to becoming organelles housed in special vesicles in the clam cells—completely necessary for the clam's survival, with the bacteria now incapable of living independently apart from the clams. This case parallels the events that led to mitochondria and chloroplasts, though certainly much more recently since clams are far more recent additions to the Earth than protists.

Another example involves bacteria of the genus *Buchnera* and aphids—those tiny insects that are considered pests on crops and many ornamental plants around one's home. Aphids are insects that suck plant sap exclusively as their diet. Plant sap is typically low in protein and does not contain all of the 20 necessary amino acids that animals require. *Buchnera* is now a permanent endosymbiont of the aphids and is apparently found nowhere else in nature. As in the case of the hydrothermal vent clams, *Buchnera* too is passed from one generation of aphids to the next in their eggs (Futuyma, 2013). The bacteria live in special host cells inside membrane-bound vesicles where they manufacture and supply up to 10 amino acids essential to the aphids. *Buchnera* too has lost most of the genes it would require if it were still free living, so the two organisms have apparently been tied together in this endosymbiotic relationship for a very long time—estimated at 150 million years minimum. Today there are close to 5,000 species of aphids, with most apparently utilizing this same endosymbiotic arrangement. In this case, the endosymbiosis has seemingly allowed a large group of species to come into existence and thrive. For reference, there are about the same number of aphid species as mammal species, so this is a significant part of species diversity on our planet.

Other similar cases of endosymbiosis involving other insects are regularly being discovered. Several other diverse groups of sap-feeding insects, including leafhoppers, plan-thoppers, spittlebugs, and cicadas, have independently formed similar endosymbiotic relationships with various genera of bacteria. Some of these insect groups have even evolved two or more joint endosymbiotic partners where each symbiont fills a different nutritional or metabolic need. Some of the blood-feeding insects like the tsetse flies have likewise evolved with a bacterial endosymbiont (*Wigglesworthia*) that provides vitamins not found in their diet of blood. As in the case of *Buchnera* and aphids, these additional examples of endosymbiosis are each believed to be many millions of years old.

There are undoubtedly a great many more undiscovered cases of this type of complete endosymbiosis where the symbiont no longer lives apart from its host and is passed/reproduced vertically within the host line of descent. Wherever this phenomenon has occurred, two separate species have effectively fused into one, even wherein the more recent mergers we can still identify the endosymbiont with a family, genus, and species name, usually with known relatives that are still free-living. An obvious question is: When do such vertically inherited endosymbionts "officially" become organelles rather than "endosymbiotic prokaryotes?" It would seem that the answer might logically be when the endosymbiont can no longer survive separately in nature, and all of the host population contains the endosymbiont. This condition would result when the symbiont had lost enough of its original genes so as to be completely dependent on the host cell/organism for its survival and reproduction. Some would also add that many of the genes originally found in the symbiont have been relocated into the nucleus of the host eukaryotic cell. By these suggested criteria, *Buchnera* is, and has apparently long been, an organelle in aphid tissues.

Most of the history of evolution consists of splitting (speciation), as is often diagramed in cladograms and other tree-based representations, but in these cases of fusing endosymbioses, there has been a fusion of life forms—and typically of two vastly unrelated lines of descent. This still surprising point is discussed and illustrated further in Chapter 13.

A much larger number of close symbioses exist where the two partners are coevolved in some respects, but the endosymbiont still lives freely in nature and must be reacquired by every host generation. The some 15,000 species of lichens are good examples of this. Lichens are literally combined organisms consisting of a fungus and either a green alga or a cyanobacterium. The genus and species name for any lichen refer to the fungal partner, with the photosynthetic symbiont having its own separate genus and species name. There is more fungal diversity in lichens than algal or cyanobacteria diversity—in other words, the same alga may be the typical symbiont with a great many fungal species, and the same goes for the cyanobacteria. The common and widespread cyanobacteria *Nostoc*, for example, is a widespread partner in many lichen species. Most lichens reproduce both sexually and asexually. In most cases of asexual reproduction, the asexual propagule contains both the fungi and the photosymbiont, but in sexual reproduction the sexually reproduced fungal spores must germinate and "find" a suitable alga or cyanobacteria, so apparently some element of luck determines which sexually produced spores will be able to locate the required photosymbiont. Some lichen species are not known to reproduce sexually, so the photosymbiont in these cases would be mainly reproduced vertically with its host, and may over time be expected to diversify from its free-living relatives, unless those same lichens also continue to incorporate the free-living forms as well. It is believed that lichens are a polyphyletic group—that is, there have been several independent acquisitions of "lichenization" by various fungal lines at various times in the past, so lichens cannot be said to form a single true clade.

The hermatypic corals and their mutualistic zooxanthellae (typically photosynthetic dinoflagellates) are another textbook example of symbiosis in which the photosynthetic zooxanthellae are reacquired by each new generation of coral polyps. The flatworm *Convoluta roscoffensis* and its algal endosymbionts in the genus *Tetraselmis* constitute another classic example. In the corals, the photosynthetic zooxanthellae share some of their produced carbohydrates, which can account for 50–90% of the fuel needs of the coral. In the flatworm *Convoluta*, once enough of the algae have been ingested and incorporated into its tissues, the flatworm's digestive tract actually becomes reduced and nonfunctional with the *Tetraselmis* algae then meeting 100% of the worm's needs. The worm in effect becomes an autotrophic animal due to its acquired store of photo-synthetic endosymbionts.

There are a great many more known examples of elegant symbiotic relationships that span a wide range of biodiver-sity, certainly enough to fill several volumes. Recall the legumes and the symbiotic nitrogen-fixing bacteria in their root nodules, or most trees and their mycorrhizal fungi—two other major examples of these relationships. Consider also that insects and nematodes together constitute over half of all described species, and there are most probably prokaryotes associated with each of these species in some form of symbiotic relationship—like the well-known case of termites and their special gut flora of both prokaryote and protist symbionts. To date, scientists have probably discovered only the tip of what logically must be a huge iceberg of symbiotic arrangements involving the bulk of life on Earth.

And the relationships do not have to be mutually beneficial as in these examples of mutualisms. Around two-thirds of the Earth's species are believed to be para-sites, and parasitism has evolved independently in a great many lines of descent (vampire bats, fleas, flatworms, nematodes, several protists, several plants, etc.). Since all tapeworms (about 1,100 species) arose from a common tapeworm ancestor, as did all fleas (about 2,000 known species), some of these parasitic groups have contributed significantly to the Earth's total biodiversity.

Another evolutionary aspect of parasitism is that parasites and infections have been the driving force in the evolution of immune mechanisms and systems in an array of organisms. This is best known in animals, where leucocytes, interferons, interleukins, antibodies, complement, and many more effector molecules and cells have evolved over the long coevolutionary history of the various hosts and parasites. In essence, all of these specialized cells and molecules exist because of a long evolutionary history of interactions with parasites. Parasitism has also spawned numerous host behaviors such as grooming in primates and cleaning mutualisms in some marine fish. William Hamilton proposed the now somewhat accepted idea that the initial evolution of sex was in large part to allow host organisms the ability to more quickly evolve resistance to the countless parasites they were exposed to. Spanning a vast range of particulars, symbiosis has undoubtedly had at least some effect on the evolution of almost every life form on the planet, and it of course continues to be a vibrant factor in the ongoing story of evolution.

Chapter 13

Speciation

...he (Darwin) concluded that one colonizing South American mockingbird species had produced three different new species of mockingbirds on different islands in the Galapagos.

Ernst Mayr

As discussed in Chapter 1, Darwin's two most important contributions to our understanding of evolution were descent with modification (mainly due to natural selection) and common descent (species have multiplied over time from common ancestors). Common descent was not accepted by most evolutionists of the early 1800s (like Lamarck), but since that time it has become a bedrock fundamental of modern evolutionary thinking. A significant amount of the work and thinking that evolutionary biologists engage in centers around common descent and its ramifications, such as homology (Chapter 15) and phylogeny (Chapter 20). Cladogenesis refers to the origin of new species through a process whereby one species splits into two or more separate species. Cladogenesis is held to be the process through which the vast majority of the Earth's species originated. Later in this chapter we will examine alternative ways in which new species can come into existence, but cladogenetic splitting still appears to be by far the more common method of new species formation.

Successive occurrences of speciation in a line of descent will eventually lead to several species, all related through a common ancestor. The resulting species (extinct and extant), including the common ancestor, are referred to

Evolution. http://dx.doi.org/10.1016/B978-0-12-800348-0.00013-4
Copyright © 2014 Elsevier Inc. All rights reserved.

together as a clade. Birds are a clade since evidence indicates that all birds, both living and extinct, descended from one common ancestor. Four other examples of clades would be rattlesnakes, flowering plants, bats, and beetles. Speciation by splitting can be said to occur mainly in two fundamental ways, known as allopatric and sympatric speciation, though there are other subcategories that we will take up later in the chapter.

ALLOPATRIC SPECIATION

A few decades ago, most evolutionists believed that for speciation to take place, there had to occur some form of geographical isolation of some members of a species from the rest, and this isolation had to remain in effect for a great many generations to allow natural selection, genetic drift, mutation, and other mechanisms the time to act differentially on the two isolates, resulting eventually in genetic differences that at some point in time allowed the distinction of the two isolated populations as separate species. This idea was mainly championed by the great evolutionary biologist Ernst Mayr. Mayr believed that this form of speciation, which he termed allopatric speciation, was the explanation for the majority of speciation events that had ever occurred, and he might still be correct on this point. There are a great many known ways in which such geographical isolation can occur, and it might be instructive to briefly list a few of these:

- Many volcanic islands have appeared over geologic time in the oceans of the world. After they break the surface and cool, life starts arriving as spores and seeds blown on the wind or carried there by birds. Some birds will be blown to the island or find it in their normal travels. Terrestrial animals are on occasion found on drifting mats of vegetation that wash out of large rivers into the oceans. These accidental "inoculation pods" can end up washing ashore on new islands and depositing small reptiles, insects, seeds, etc. The longer an island persists, the more life it typically accumulates. All these island colonists came from mainland populations or from other islands, and once on the new island they will either die out or start adapting to the new environment of the island, which is most often different in many ways from the environment of their former home.

- Over geologic time, continental drift has slowly ripped some continents apart—as Africa was once joined to South America. The rifting of these two landmasses, which slowly formed the southern Atlantic Ocean, separated many thousands of animal and plant populations into two separate ranges with little or no chance for future interbreeding, and with increasing environmental differences as the distance between these now separate continents increased (and is still increasing).

- Ocean levels have risen and fallen many times over geologic time. Most of the area of Texas, Oklahoma, and Kansas was once covered by a shallow sea when ocean levels were higher. Land between Siberia and Alaska was once exposed when sea levels were lower. In the latter example, some species were isolated from each other when the ocean later rose, resulting in several miles of ocean between Siberia and Alaska. A similar occurrence occurred long ago when the thin connection between North and South America was submerged (Panama and neighboring countries). Also, for freshwater fish populations, when the ocean level rises past a branch connection of two separate river forks, the two forks of the river then become separate rivers, which may remain separate for millions of years, during which time evolution may well proceed differently in the now-isolated populations of several fish species (freshwater fish would not typically swim out into saltwater to find the other river, which was once a part of the one former river system).

- Mountain ranges can arise due to geologic uplift. When the continental plate carrying what is now India met the southern edge of the Asian plate, the slow and still ongoing collision of the two landmasses pushed up the

Himalayan mountain chain—today the world's highest. For a time before the uplift, many species would have inhabited the areas both north and south of the uplift zone, but as time proceeded and the uplift rose, most species were eventually cut off into a group north of the uplift and a group south of the uplift—with no chance of continued inbreeding between the two separate populations. Many such uplift events on most continents have occurred over geologic time due to various geological processes, and they all separated a great many species into two or more isolated groups.

- Though this example may seem far-fetched, over geologic time it has undoubted occurred numerous times. When tornados move over water they suck up water from the lake or river along with fish, zooplankton, snails, frogs, turtles, and other aquatic life. When the tornado dissipates, as they all do, the life forms now high in the atmosphere fall back to Earth, often still alive and many miles from their source. Odd reports of fish or frogs "raining" from the sky are true in some cases due to this very phenomenon. In some even more rare situations, the raining fish, frogs, snails, and such fall back into lakes or rivers other than the one they were lifted from. It would only take two or three individuals to start a new population of a fish or snail species in the new lake, and even if separated by only a few miles, the two lakes would differ in some ways, such that over geologic time this new population would evolve into a new species distinct from its parent species. Less severe weather such as large storms can blow birds and insects far from their native ranges and sometimes deposit them in new areas that are geographically isolated from their former range—obviously so in the case of animals being blown to islands, isolated valleys, plateaus, etc.
- Caves, like islands, are well-known laboratories of allopatric speciation. When caves enlarge and occasionally open to the surface, the chance for entrance of surface-dwelling fish, salamanders, arthropods, and worms often

results in some individuals "falling in" to either die in the cave (probably the more common result) or to survive and start adapting their small populations to this drastically different environment. At least hundreds of insular animal species from around the world are known that are cave-dwelling and cave-adapted. These often include species that have lost or reduced their eyes, their skin pigments, and their body volume to aid them to live in this dark world where every calorie must be conserved because of the scarcity of available food.

- Rivers are barriers to many small animals, and sometimes they define one edge of the range a species. Rivers can and do alter their course over geologic time, and in the process can isolate a few individuals from their parent population—on the other side of the river. If such a founder population survives and thrives, it may start slowly evolving differences from the parent population from which it is now cut off.

- In some places like Hawaii, lava flows can divide one area of forest from another. Even a hardened lava flow of only a few yards wide can serve as an effective barrier to some amphibians, lizards, and insects, which can then speciate during their isolation—as it will take many years for the lava to erode and allow the forest to reestablish itself over that area.

These examples do not exhaust the possible ways in which a species can become divided geographically. There are others, but hopefully the previous examples will illustrate the variety of ways in which physical separation can and does occur. Once separated into two isolated breeding populations, speciation would be expected to occur eventually due to three factors: (1) genetic drift acting differently in each population, (2) neutral or beneficial mutations occurring and spreading within each isolated population (but of course these would most likely be different mutations in each group), and finally (3) natural selection, which would probably be acting differently in the slightly different to drastically different environments of the two populations. Whatever the source of genomic and gene pool change, these changes would be contained and would accumulate within each of the two isolated populations—so leading to greater and greater genetic divergence over time.

For a long time, most biologists accepted and followed Ernst Mayr's lead in accepting that almost all speciation was achieved through allopatric speciation—it was just so logical and easy to envision—not to mention the many known examples that gave every appearance of having come about in this way. The tortoises and iguanas that arrived, became isolated, and diverged on the islands of the Galapagos are two well-known examples, as are the many cave-adapted species of fish.

SYMPATRIC SPECIATION

Another possibility that was recognized, but initially with little supporting evidence, was that of sympatric speciation—speciation without geographical separation of population. This idea carried with it the great problem of how organisms would start diversifying when the two incipient species were still mingling with each other in the same ecological space, and therefore continuing to be exposed to each other as potential mates. What would drive the separation and the assortative mating in such a situation?

There are now at least a few examples of what definitely appears to be sympatric speciation—either completed or in progress. One popular example of sympatric speciation seemingly nearing completion is that of the hawthorn fly *Rhagoletis pomonella* of the northeastern United States. This small fruit fly had coevolved with hawthorn trees in that males court females on the hawthorn fruits, mating occurs on the hawthorn fruits, the mated females lay their eggs on hawthorn fruits, and the hatched larvae then feed on the hawthorn fruits. Then apple trees were introduced into this region from the 1600s. Apparently sometime after that a few of the hawthorn flies took to venturing over to this new, larger and juicer fruit, which was red just like the hawthorn fruit. For a great many years now the ranges of hawthorn and apple trees have overlapped (they occur in sympatry), yet some of the *Rhagoletis* flies seem to now be specialized on apples, and are now showing slight morphological differences from those that continue their association with hawthorns. It is also found that flies that develop in apples will as adults prefer apple trees to hawthorns, again locating mates there, mating there, and having their offspring deposited into apples (Schilthuizen, 2001). This ongoing trend of course only strengthens any inherited preferences and behaviors in the flies and adds to this divergence.

Currently, it is judged that the two types have not speciated completely because the two fly types still interbreed readily in laboratory conditions and produce viable offspring. In nature, though, they seem to largely associate and interbreed only with their own "group". Since fruit flies can go through several generations in a year, their evolution can occur rapidly. If the trend continues, it will likely not be far in the future when we can conclude that this speciation story is complete—that two distinct species of *Rhagoletis* have evolved from one.

Another now well-known set of examples involves the freshwater fish known as cichlids. Both in Africa and in Central America there are lakes holding several species of cichlids that are genetically closely related—more so within any one lake than to the cichlids of any surrounding lakes. The lakes in question are also known to be relatively young lakes in geologic time—only a few thousand years old. In many of these cases, it appears that speciation of a colonizing species has occurred within the same lake without geological isolation, and some of these speciation events seem to have occurred within as little as a few hundred years (Schilthuizen, 2001). Sympatric speciation seems to be the explanation, perhaps due to developed preferences for depth, feeding sites, diet, etc. Differential sexual selection is also thought to have contributed to some of these sympatric speciation events.

Sympatric speciation is usually due to what is known as disruptive selection, a type of natural selection in which two distinct variations in a trait are both favored as opposed to an intermediate form of the trait (see Chapter 2). In the case of the hawthorn/apple flies, flies that associate with either hawthorn or apple trees are seemingly favored, but not those that choose either tree randomly. In the case of some of those cichlid examples, those that preferred a shallow or a deep depth were favored, but not those that preferred an intermediate depth. Undoubtedly many more cases of sympatric speciation will be discovered in due time.

PARAPATRIC SPECIATION

In yet another variation of divergence, in and around White Sands National Monument in New Mexico, there live two colored forms of lesser earless lizards (*Holbrookia maculata*). Those that live out on the white sands have evolved a whitish skin, allowing them to be camouflaged against the white background. Those lizards living in the larger surrounding desert terrain have a mottled brown coloration with traces of green and orange—a coloration that allows them to more easily blend into their desert background. Geologically, the White Sands Formation is estimated to be only about 7,000 years old, suggesting that the white lizards are the newly derived form, with their color serving as an adaptation for living on this new white substrate (Conrod and Rosenblum, 2008). In tests of preference, white males preferred to mate with white females as opposed to the brown females. Though these two color morphs are not yet judged to be two separate species, we can see here disruptive selection working across an environmental boundary. The lizards could easily move between the white sands and the normal desert substrate, but they seem to prefer to stay on one or the other and chose mates of their own color morph. If this preference eventually leads to complete speciation, this would be a case of parapatric speciation—speciation across an environmental feature that does not prevent movement of the individuals (as in allopatric speciation), but which may influence selection in different ways on the two sides of the junction.

SPECIATION BY POLYPLOIDY

As mentioned in Chapter 5, the most extensive kind of chromosomal mutation is a duplication of the entire genome; known as polyploidy. This usually occurs due to a failure of chromosome separation in meiosis such that some of the gametes come out diploid (2N) rather than the expected haploid (N). If these 2N gametes are fertilized by normal N gametes, the resulting embryo will be triploid or 3N. In many organisms the 3N combination is fatal. Human 3N embryos almost never come to term, and if they do they die shortly after birth. If not fatal, the 3N individual is almost always sterile because this odd number of chromosome replicates tends to cause insurmountable problems in meiosis such that viable gametes are extremely unlikely to occur. It is not well understood why having additional copies of chromosomes or genes is often harmful in some species, but it is. Recall that individuals with Down syndrome have only one additional chromosome—an extra copy of chromosome 21, and this confers a range of mental and developmental problems.

For reasons still incompletely known, plants seem to tolerate a 3N condition such that some 3N plants can survive and grow to adulthood. There are two ways in which polyploidy can be perpetuated. More commonly in plants, some species reproduce mainly by asexual means, and a 3N chromosome number poses no grave problems for asexual reproduction accomplished through mitotic cell divisions. If asexual reproduction of this type is easily continued, then a new species can be created that is limited to asexual perpetuation.

On the other hand, if the original 2N gamete were to be fertilized by another "abnormal" 2N gamete, then the resulting embryo would be 4N, which would present no major problems to future sexual reproduction involving meiosis. However, sexual reproduction would then be limited to crossing with other 4N individuals, or with one's self—and facultative self-fertilization is not at all uncommon in many kinds of plants, and even a few animals. In any of these variations of autopolyploidy (polyploidy involving a single species origin), new species may be instantly created that will be self-perpetuating if they suffer no fitness defects due to the polyploidy. The cultivated species of potatoes and bananas are the result of autopolyploidy.

SPECIATION WITHOUT CLADOGENESIS

Though cladogenesis explains most of the increase in species diversity over evolutionary time (including those that arise through autopolyploidy), there are other ways in which new species can arise. One of these is through endosymbiosis, with the most prominent example being the one discussed in Chapter 12 that led to the origin of the first eukaryotic cells—not only a new species, but a new domain was generated by this symbiotic union. At least for a time after such an endosymbiotic union, most members of the two interacting species would continue to exist, while those that entered the endosymbiotic relationship become something new—a new species in at least some cases. A diagram of this type of speciation might be described as a fusion rather than a cladistic splitting, as illustrated in Figure 13.1.

Another noncladistic method of species formation that would also fit Figure 13.1 is that involving the hybridization of two species accompanied by polyploidy—known as allopolyploidy. In Washington state, *Tragopogon mirus* is a relatively new polyploidy species of flowering herbaceous plant that resulted from the hybridization of two previous species: *Tragopogon dubius* and *Tragopogon porrifolius* (Schilthuizen, 2001). These two original species each have 12 chromosomes as the 2N number in their cells, while the new hybrid has 24. The details of exactly how the chromosome number doubled need not concern us, except to

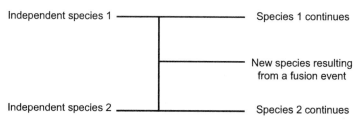

FIGURE 13.1   Fusion speciation. "Instant" speciation through fusion of some members or gametes of two separate species. The two species involved in the fusion may continue largely unaffected after the fusion event.

say that by some failure of meiosis in the original gametes leading to the first hybrid, or some failure of mitosis in a resulting zygote, the number was doubled to 24. *Tragopogon mirus* is now an established species, which being 4N compared to the two parent species can reproduce sexually within its members. An example like this is obviously yet another variation on sympatric speciation (since the two parent species lived in the same habitat); however, this would be a fusion rather than a cladistic case of sympatric speciation.

Again, plants are rife with past polyploidy events, both autopolyploid and allopolyploid, and thus much of plant species diversity stems from speciations (at least in clade ancestors) involving polyploidy. While polyploidy of any origin is less common in animals, there are examples of its occurrence in amphibians, fish, crustaceans, and other animal groups. We are still in the process of deciphering how widespread polyploidy has been in the history of life on Earth. To sum up this chapter, there are many possible pathways toward speciation, not all of which with their attendant variations could be covered in a single chapter. Whole books have addressed the topic of speciation, with in-depth discussions and multiple examples. Needless to say, our modern understanding of "the origin of species" now is far more complete than anything present in Darwin's understanding, and it will undoubtedly only grow more so as evolutionary biology continues to evolve.

Chapter 14

Micro- and Macroevolution

In some cases, macroevolution will likely be extrapolatable from microevolution; in other cases macroevolution will likely not be extrapolatable from microevolution.

Mark Ridley

It was back in the early 1970s that I first came across the terms microevolution and macroevolution in a book in the library. I was a graduate student at the time with a BS in biology. When I first read a definition of macroevolution that defined it as having to do with the evolution of the higher taxa, and of major morphological, physiological, or behavioral transitions in the past—I was confused. I wondered how this could be different in any way from the normal day-to-day evolutionary change in species that would simply accumulate through time and diversification to result in those things macroevolution was said to be concerned with. In short, regular evolutionary change due to natural selection, genetic drift, and neutral evolution (microevolution) accumulated over time would result in the diversification of the higher taxa and any "significant" evolutionary transitions in morphology, physiology, or behavior.

But back in the 1970s, symbiosis, horizontal gene transfers, transposable elements, and the power of developmental alterations were not yet well understood or well accepted as additional and important mechanisms of evolutionary change. Even with a modern understanding that incorporates these additional and different mechanisms, it still seems

Evolution. http://dx.doi.org/10.1016/B978-0-12-800348-0.00014-6
Copyright © 2014 Elsevier Inc. All rights reserved.

logical that most of so-called macroevolutionary change is the result of "normal" microevolutionary change accumulated over time. There is no known reason to suspect that the roughly 40 species of dolphins distributed over several genera came to exist through anything other than accumulated microevolutionary change and cladogenesis, plus of course the extinction of past intermediate and transitional forms. The same goes for the 35 or so species of rattlesnakes, or the 14 species of Galapagos finches, or the recent split that gave rise to the chimpanzee and the bonobo.

Some brief definitions of macroevolution simply say "evolution above the species level". If you consider that particular wording, and you know something about evolution, you would likely come away confused at this wording. All evolution occurs within populations and species, and that type of evolution is easily conceived as the type of change that resulted in most of the higher taxa, again accumulated over long spans of time.

Our modern knowledge of speciation is still incomplete, but we do know that speciation can sometimes occur rapidly owing to things like differential sexual selection or habitat selection in populations, whether they be in allopatry or in sympatry (see Chapter 13). These changes are certainly cases of microevolution in the majority of cases. Even what humans have labeled "major transitions" were probably the result of slow, regular microevolutionary change. Though we will never know this with certainty, the transition from the already "limbed" fish-grade animals like *Acanthostega* to an early amphibian-grade animal like perhaps *Hynerpeton* (we will never know whether *Acanthostega* was in fact a direct ancestor of *Hynerpeton*) most likely involved only a little gradual genomic change that probably would not qualify as macroevolutionary change.

It remains more than a little subjective as to exactly where to draw a line between microevolutionary change and macroevolutionary events, and the two are not mutually exclusive in all cases. The acquisition of feathers by some of the theropod dinosaurs and the later transition of one of their lines into "birds" is used frequently as an example of a macroevolutionary transition, but the many new fossil finds from China indicate a more gradual and stepwise process. The first feathers seem to have been small filament-like structures unlike the larger expanded feathers found in later forms. These early feathers did not distinguish the animals bearing them significantly from their close relatives of the time. Even powered flight was probably acquired in a stepwise process of evolution involving gliding stages followed by weak flight that certainly lacked the speed and dexterity of a modern swallow or swift. It is easy and subjectively tempting to label some evolutionary changes as macroevolutionary when viewed only dimly through an incompletely understood past consisting of many millions of years.

Today, however, there are a few known examples of evolutionary change that arguably do seem to qualify as "macroevolutionary". The most obvious of these are the cases of complete endosymbiosis of one organism inside another, with the best-known example being the eventual origin of the eukaryotic cell, wherein mitochondria evolved from a line of purple alpha-proteobacteria that became endosymbionts inside what is believed to have been an archaeal-line cell some 2,000,000,000 years ago. This event, which gave rise to a combined organism (especially the combined genomic result), was certainly not microevolutionary change, nor was the later similar event some 1,200,000,000 years ago when a cyanobacteria likewise became an endosymbiont of a eukaryotic cell to form the world's first alga. These and other similar macroevolutionary events are discussed further in Chapter 12.

Another clear case of macroevolutionary change occurs when a new species is suddenly created through polyploidy. Polyploidy refers to the situation where a genome is accidentally doubled, tripled, quadrupled, etc. This can come about through a few different pathways. It often occurs through a failure of chromosome separation in meiosis such that some gametes come up diploid (2N) rather than the normal haploid (N). If a diploid egg is fertilized by a normal haploid sperm, a triploid embryo would result. In some species, this is not fatal and a 3N organism develops. These triploids (3N), however, are often sterile because triploidy creates grave problems for the process of meiosis such that only nonviable eggs or sperm are produced by the triploid. Even more rarely, the egg and sperm from different individuals have both had a chromosome separation failure in meiosis such that both sperm and egg are diploid. If these unique diploid gametes meet and join, the resulting fertilization gives rise to a 4N zygote with double the chromosome number of the two parents. Chromosome number can also rarely double in an originally 2N zygote, resulting in a 4N zygote.

If a 4N zygote is viable, it may survive and produce viable 2N gametes (no grave challenges in meiosis with an even chromosome count) that are now able to successfully fertilize only other 2N gametes, giving rise to yet more 4N individuals. Many plants, and some monoecious animals, are capable of self-fertilization. If a 4N individual is generated, it may self-fertilize and create yet another generation of even more 4N individuals. If this process can keep repeating itself, a new species of plant or animal can be created within a couple of generations, obviously with a new chromosome number, and often with a few new morphological features that result in some unknown way from the enlarged genome. It is estimated that most plant species today have one or more polyploidy events back in their ancestry. The situation for animals is less well known, but polyploidy clearly has occurred in some arthropods, fish, salamanders, and other groups. This unique mechanism of rapid species creation would seem to qualify as a form of macroevolution.

In prokaryotes, the process of horizontal gene transfer (HGT—discussed in Chapter 7) by plasmids can in some cases arguably result in macroevolutionary change. Some plasmids carry several genes into the receiving bacterial or archaeal cell that instantly give that cell new properties such as chemical resistance to certain antibiotics, the ability to become a virulent parasite in some host organisms, or the ability to metabolize different molecules in its environment. Some of the various photosynthetic bacteria are now believed to have acquired the genes that support photosynthesis through a small number of HGT events—obviously resulting in a major lifestyle change. Such rapid acquisition of important properties through newly acquired genes certainly seems to qualify as macroevolution in the vast and ubiquitous realm of microbes.

The term macroevolution might also arguably be applied to cases like that of the axolotl described in Chapter 11. Though this transformation of the tiger salamander into a distinctly different sort of aquatic salamander most likely involved only one or two mutations that downregulated thyroxine production, the resulting organism, the axolotl, seems distinctly different from the tiger salamander and does constitute a different species. Any mechanism that rapidly generates a new species could be arguably termed macroevolution. Yet another possible example of macroevolutionary change could be when a parasitic species either adds or loses a host from its life cycle. There are various hypotheses concerning particular parasitic genera, proposing in some cases the rapid addition of a new host to the life cycle, or the loss of a former necessary host. Assuming these changes occur and quickly become established in a line of parasites, such significant life cycle alterations might also arguably be viewed as macroevolutionary changes. These examples probably do not exhaust the possibilities, and there may be at least a few more varieties of rapid and significant change (some most likely still unknown) that could reasonably fall under the label of macroevolutionary events.

Chapter 15

Homology

Common sense says that any particular complex organ, requiring many different genes for its construction, is likely to evolve only once; so that all creatures with such an organ must indeed have inherited it from the same ancestor.

Colin Tudge

...it is plain that all the myriad forms of myosin evolved from a common ancestor.

Nick Lane

If someone fully understands the meaning of the single term "homology", she by default has a fairly good understanding of evolution as a whole. Evolutionary change over time was well accepted by several scientists prior to Darwin, with Jean-Baptiste Lamarck being the best known example. Lamarck, however, along with several others, did not comprehend or accept the idea of common descent, which Darwin argued so forcefully—and which has since become an essential component of our evolutionary understanding.

"Anagenesis", or phyletic evolution, as accepted by Lamarck would have no need of the term homology. Anagenesis is defined as evolutionary change within a single line with no speciation or splitting of that line. Anagenesis undoubtedly does occur in some lines of descent over shorter spans of geologic time, though concrete examples are hard to prove because of the incompleteness of the fossil record (many sister species are undoubtedly still undiscovered). Lamarck accepted only anagenesis, envisioning all species changing or evolving over time—but without any one species

Evolution. http://dx.doi.org/10.1016/B978-0-12-800348-0.00015-8
Copyright © 2014 Elsevier Inc. All rights reserved.

ever splitting into two or more species, no matter how long successive changes had been occurring. Only "cladogenesis" (Chapter 13), or the splitting of one species into two species (and eventually into millions of species since the origin of life), has need of the term homology—literally structures or features found in two or more species which were inherited from a common ancestor—or as we say, through common descent. Of course, it is understood that what is actually inherited are the genes that code for those structures, not the structures themselves.

All birds have feathers. The feathers of all birds are homologous because they all inherited feathers from a common ancestor in the theropod dinosaur group. Today the feathered dinosaurs are all extinct, unless you define birds themselves as extant feathered dinosaurs—which some biologists do. To the best of our understanding, all birds that evolved from that ancestor, extinct and living, had feathers—therefore the feathers of any two species of bird are homologous. Yes, the feathers of modern birds are extremely diverse from species to species (as well as diverse even on some individual birds), but all are descended/inherited from that common ancestor. This is what it means to say that structures are homologous.

Another good example would be the femur bone of all legged tetrapods (amphibians, reptiles, birds, and mammals). Tetrapods that have hind legs have femurs. Even some whales, pythons, and other species that have almost lost their hind leg bones still have vestigial femurs. The first probable land vertebrate ancestors like *Ichthyostega* had a femur, and all modern tetrapods share a femur through common descent from a common ancestor that was at least a close relative of *Ichthyostega*.

Examples of homologies exist by the thousand—from the level of large structures like feathers and specific bones, down to the level of genes like the gene for cytochrome C (shared by all animals), myosin (chapter lead quote), and other molecules. The near-universal consistency of the genetic code is considered by biologists to be the supreme example of an extremely old homology shared by all living organisms. It is true that a few prokaryotes have a few derived differences in the code, just as femur bones differ between frogs and tigers, but that is only to be expected considering the span of time that this code has been in existence—probably more than 3.5 billion years. The protein actin is another homology shared by all animals and a great many protists. It is one of the most common proteins on the Earth, making up a large percentage of animal muscle and serving several other roles in assorted organisms. All animals inherited actin (again, more precisely—the gene that codes for actin) from a common protist ancestor, so it is homologous in all animals.

Behaviors too can be homologous since most behavior is innate—resulting from genes coding for nervous systems hardwired for the performance of certain behaviors. Considering that the vast majority of behavior occurs in protists, insects, spiders, worms, jellyfish, and fish, learning plays a minor role in the great majority of behavior. The flight behavior of all butterfly species is therefore homologous, as are the pulsating swimming contractions of all Cnidarian jellyfish. Even in the land vertebrates there are countless examples, such as flight behavior in the thousand-plus species of bats, rattle rattling in all species of rattlesnakes, nectar feeding in hummingbirds, and countless others. If all evidence suggests that several modern species in a group plus all intermediate ancestors back to and including a common ancestor had the behavior in question, it is then a homologous behavior in that group. Admittedly, for behavioral examples one has to apply some logic and parsimony analysis to say that particular behaviors are homologous, though correlations with morphological and molecular homologies can greatly strengthen such conclusions.

The concept of homologous structures, molecules, and behaviors is often contrasted to structures that are similar—and often serve similar roles, but which are not homologous through common descent. A great example of structures that are similar but obviously not homologous would be the prehensile tails of seahorses and some monkeys, like the spider monkey. The seahorse is a bony fish, and the vast majority of bony fish do not have prehensile tails—rather they have the "normal" finned swimming tail used in forward propulsion. Seahorses evolved from something like the modern pipefish (a member of the same family), which has a more fish-like tail (though tails are more slender and reduced in pipefish than in most fish). Seahorses use their prehensile tails to spend a lot of their time hanging onto sea fans, branching corals, sea grasses, and other holdfasts in the marine environment where they can wait for assorted zooplankton to swim by—one of their favored food types. Monkeys are among the very few mammals that have prehensile tails, and they surely evolved these tails as primates rather than inheriting them from earlier mammalian ancestors. There exists a rich fossil record of early mammals, and their remains do not indicate animals with prehensile tails, or the need for prehensile tails.

Prehensile tails evolved then in one small group of fish—the seahorses, and they also evolved independently within a small order of mammals—the primates. Seahorses and spider monkeys therefore do not share their prehensile tails through common ancestry, rather each group derived this feature independently. In the jargon of biologists—their evolution converged on prehensile tails as adaptive features—one in the sea and the other in the treetops. Such shared traits are said to be analogous or convergent. Prehensile tails evolved independently in a few other groups as well, like the chameleon lizards of Africa (Figure 15.1). Prehensile tails are but one of a great many examples of convergent evolution, where two or more different species have independently evolved some feature that, though similar in many ways, is nevertheless not shared through (or because of) common ancestry.

(A)

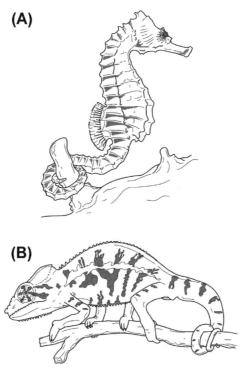

(B)

FIGURE 15.1 The convergence of prehensile tails in (A) a seahorse and (B) a chameleon. *(Figure by Jeff Dixon.)*

Here is a short list of a few other examples of convergences (analogous traits) in two groups that do not share the trait in question through common ancestry:

- Suckers—a leech and an octopus.
- General body shape—dogs and the extinct thylacine (an Australian marsupial). See Figure 15.2.
- Loss of the digestive tract—tapeworms and acanthocephalans. Both are parasites of the digestive tracts of vertebrates, but are very distantly related. In fact, recent genomic evidence indicates that acanthocephalans are probably highly derived rotifers, most of which have fully functional digestive tracts.
- Social behavior and caste systems in termites and ants. Termites are not close relatives of ants. Ants evolved from wasps, while termites evolved from roach-like ancestors that lacked such social systems.

FIGURE 15.2   A thylacine. A recently extinct marsupial of Australia and Tasmania showing a strong convergence of form with dogs, which are distantly related placental mammals. *(Figure by Jeff Dixon.)*

- Loss of limbs in snakes and in some lizard groups like the glass lizards. Though both are reptiles, each of these groups lost their limbs independently and do not share their limbless condition through common descent.
- Loss of flight ability in the dodo and kiwi birds (and several others).
- Loss of eyes—some cave-dwelling salamanders and some cave-dwelling fish.
- Wings in bats and wings in birds.

In this last example, there exists the possibility for confusion because within the wings of bats and birds are some similar bones and muscles. In fact, the forelimbs of bats and birds are homologous—as are most of their bones and muscles. But they are not homologous *as wings*. The "wingedness" of birds evolved in some of the small feathered theropod dinosaurs, while the wings of bats evolved independently and much more recently in a once small group of mammals. Bats are the only flying mammals ever to evolve, and though we are still not exactly sure what kind of mammal the first bats evolved from, they or their immediate ancestors were not flying animals and so lacked wings.

In most, but certainly not all, cases of convergence, it is typically just the one convergent trait that is obviously similar between the two species involved. A monkey and a seahorse do not share many other specific traits in common—what with one being a marine fish and the other being a primate. The same applies to birds and bats in overall structure, and even to dogs and the thylacine if you are an expert in mammalian anatomy. Though the outward overall shape of a dog and a thylacine are indeed remarkably similar, the thylacine is a marsupial with a fundamentally different reproductive physiology than that of dogs. The skeletal structure of the two also shows a great many fundamental differences—especially their skull structure.

Again, behaviors can be convergent as well, whether innate or learned. Vampire bats evolved to feed on blood, but convergently so have ticks, mosquitoes, leeches, and hookworms. These groups are certainly not all blood feeders due to homology or common ancestry. Ticks are arachnids and most arachnids do not feed on blood, leeches are a derived group of Annelid worms most of which are not blood feeders, and hookworms are nematode worms, and most nematodes are not blood feeders. Hopefully this is all perfectly clear, but just to add—there are about a thousand species of bats, and only the vampire bat is a blood feeder. Everything about this situation tells us that the vampire bat is a recent addition to the bat horde, and its blood feeding is certainly a recently derived and isolated appearance of this feeding type (among bats, that is).

The living world is filled with countless homologies among closely related species, and even many distant ones. To repeat, these range from behaviors, to morphologies, to molecules, and to genes. As for convergences, they are far fewer in number, even though some are striking in their similarity. When a biologist is working out the probable evolutionary relationships of species, clear homologies are invaluable in grouping species into related groups. Since vertebrae are recognized as homologous in all animals having vertebrae, we recognize vertebrates as a clade of thousands of related species ranging from fish to mammals. Since mammary glands are almost certainly homologous, we are able to have confidence that mammals also form a true clade of related species—a smaller one than vertebrates, of which they are a sub-clade. Mistakes are possible when convergences are mistakenly thought to be homologies. This is unlikely to occur in two groups as distinct as seahorses and monkeys, but could throw off a researcher when the convergence is more significant and involves the whole body form, as in the case of dogs and the extinct marsupial thylacine.

More common are cases where a convergence causes a group of organisms to be viewed as unique and separate from a group they actually belong to phylogenetically. The general worm body form seems to be convergent in several groups. Only a few years ago the Pentastomida, a group of parasitic "worms", was recognized as a distinct animal phylum, separate from the 30-plus other animal phyla. Newer data including genomic information now indicate that these worms are actually a highly derived group of crustaceans (Lecointre and Le Guyader, 2006)! This came as quite a surprise, yet this conclusion is now well established. Though the worm form of the Pentastomids did not cause them to be grouped together with any of the other worm groups, it did keep them from being recognized as crustaceans. Though rare, convergences of this degree do exist, and they do create some difficulties in recognizing the true character and identity of certain groups of organisms. Still, biologists have made great progress in overcoming these difficulties and in determining exactly who most of Earth's inhabitants really are, though the task is of course ongoing and incomplete.

Chapter 16

Imperfection

> *Imperfect design is the mark of evolution; in fact, it's precisely what we expect from evolution.*
>
> Jerry Coyne

> *The molecular data are now in, and the scientific pictures they paint are both surprising and clear: the human genome is a Byzantine contrivance that departs dramatically from what would seem to be optimal design.*
>
> John C. Avise

In years past, it was common for biologists to be overly impressed with what they perceived as the incredibly fine-tuned nature of most organic adaptations—those resulting from the action of natural selection. Biologists still often view many adaptations as beautiful in their near-perfect execution. Examples here would be the amazing camouflage some animals have evolved—think the pygmy seahorse, flounders, and some insects like the famed peppered moth of England. Since natural selection is really a weeding-out process that on average eliminates those genomes that do not work as well as others in promoting survival and reproduction in the available environment, you might be led to think of natural selection as a "perfecting process" that over time would tend to result in near-perfectly adapted organisms. Indeed, many biologists have held this notion. But in practice, we find that natural selection is not as powerful a perfecting process as was once believed to be the case. It has slowly been realized that successful genomes and phenotypes need only be better than the competition to gain repre-

Evolution. http://dx.doi.org/10.1016/B978-0-12-800348-0.00016-X
Copyright © 2014 Elsevier Inc. All rights reserved.

sentation and dominance within populations, and they need not necessarily closely approach perfection to accomplish this. Even if perfection were to be achieved by a species, how would we recognize that quality when it is doubtful that we fully understand the complexities of life for even a single species—including our own?

Also, we now know that natural selection is but one of several processes that forge evolutionary change, and it is in fact the only one that crafts a "better" genome—a better organism (I use the term better here only to mean better than the competition at surviving and achieving genetic fitness). The other processes contributing to evolution are neutral with respect to improvements (such as genetic drift and neutral evolution) or even at odds with better design (e.g., the hanging on of vestigial genes and structures, the constraints of working from earlier designs, the largely selfish nature of transposable elements). A major factor preventing perfection is the constraint that natural selection can work only on the genetic variation currently present within each species. It cannot de novo create significant amounts of newly designed and optimal genetic material for better design features.

An analogy here might be that our modern CD players were not created by tinkering with the earlier designs for record players or tape players, which were "analog". Completely new digital technology was "intelligently designed" and incorporated into the first CD players. We would not have CD players if we were limited to a slow, step-by-step modification of the analog technology found in record players or tape players—with the added provision that any potential intermediate players would have to be able to play music and survive in the marketplace for a time. In our planned technology, designers can envision and create entirely new technologies that are not much linked to older technologies. Evolution does not have that huge advantage—it lacks this intelligent approach. As many have phrased it—evolution can only "tinker" with the gene pool at hand.

Many obvious imperfections in organisms result from this limitation of tinkering with what is already present. Perhaps the most often-mentioned such imperfection is the one that exists in our throats where our respiratory air movements cross within the pharynx with the movement of food and drink into our esophagus when we swallow. This imperfection results in thousands of choking deaths each year when a bit of meat or other solid food accidently gets past the epiglottis and becomes lodged in the upper trachea. This problem occurs often in children, and also tends to become more common and complicated in the aged—with many older individuals developing swallowing problems in which some of what they eat or drink enters the trachea or even the bronchia on a regular basis, resulting in infections and other lung and breathing problems. On a personal note, my mother probably saved my life when I was around three years old by lifting me upside down and slapping my back to dislodge a "jawbreaker" candy that had lodged in my trachea and blocked my breathing. Another problem associated with this situation is that some infections and other disease conditions in the upper digestive system can spread easily into the respiratory tract, and vice versa.

Land snails and slugs evolved lungs separately from us vertebrates, and they did so in a more logical way— they evolved a separate opening into their lung that has no connection whatsoever with their digestive tract. Land arthropods like spiders and insects likewise evolved their respiratory systems apart from their digestive systems and similarly avoided the problems faced by vertebrates. They too have no problems resulting from food interfering with their respiration in any way. Land vertebrates, however, evolved from fish that had lungs which evolved initially from side pouches off the upper esophagus. Their nostrils were at first blind pouches functioning only in olfaction, but in later lines these deepened to connect with the rear mouth cavity, which farther down the pharynx connected with those primitive lungs. This connection of the air passageway with the digestive passageway is still retained after about 400,000,000 years of evolution, even though this imperfection has undoubtedly played a role in untold millions of deaths in some terrestrial vertebrates. Have you never seen a dog get choked while wolfing down its food?

In terms of this particular vertebrate design problem, no new plan has entered the competition to gain an advantage in the arena of natural selection. A dramatically better design would require massive restructuring through developmental processes and undoubtedly the simultaneous addition, or alteration, or both of numerous genes to bring this about. In short, we land vertebrates are just stuck with this and many other imperfections that result from the constraints of past evolutionary history. It is true that in humans and some other land vertebrates the connection of respiratory and digestive tracts did allow for the possibility of vocalizations that have come to serve in several adaptive roles, but there is no reason to assume that vocalizations might not have evolved in animals having the two systems fully separated. Apparently, the land snails and insects are still getting by just fine without the need to evolve such abilities, though a few cockroaches can produce threatening hissing noises by way of their tracheal system.

Vestigial genes and structures too can be taken as examples of imperfection. If whales no longer need femurs, why do they still build the small vestigial femurs deep inside their huge bodies—connected to tiny vestigial pelvic girdles? Why waste calories and nutrients doing so? Evolution has gradually cranked down the development of whale femurs so that they stop far short of anything full-sized and functional, but femur development is undoubtedly the result of several genes and their interactions. Some of these have obviously been selected against such that femur development is severely stunted, but others continue to work resulting in the formation of these small vestigial bones. Until these genes are lost or mutated to stop femur development altogether, there will be no selection to completely rid whales of these structures, which do drain a small percentage of the whales' resources into their development.

Even if such heritable variation appears, if the advantage of having the vestigial femurs is very small, selection will be slow in ridding whales of this slight imperfection.

Standard sexual selection theory supports the idea that many imperfections in terms of survival exist owing to intersexual selection, also known as mate choice. The classic example is the huge and costly tail of the male peacock. The only known function of this long, heavy, and ornate tail is to impress the female enough to sway her to allow mating. For the male, the tail is a burden in terms of calories, nutrients, and even survival ability since it decreases his agility in escaping predators—especially his ability to take flight quickly. Now since the only thing that really counts is genetic fitness, then yes—the tail is indeed functional if females "demand" this costly burden in their males. Still, it is hard to see complete perfection in a situation where males are ultimately handicapped by the excessively escalated preferences of the females. The question is complex and can be argued either way, but certainly "perfection" is not so obvious in situations of this type, and a great many similar adaptations share this pro versus con aspect.

Another imperfection is the overly complex immune system of mammals—so many different cell types and vastly more specific chemicals, some with overlapping functions. Though no human could live very long without an immune system, this so-called system is really a hodgepodge of mechanisms that has slowly increased in complexity over evolutionary time in response to the vast and diverse array of parasites and pathogens our line has been regularly exposed to—viruses, bacteria, protists, worms, arthropods, etc. Immune responses to some parasites cause damage to the host as well as the invader, in some cases resulting in more damage than the parasite alone might have caused. The immune formation of sizable granulomas around the eggs of blood flukes in the liver results in severe liver damage. Allergies and autoimmune diseases are another set of harmful outcomes of our complex immune responses, in these cases taking inappropriate or misdirected actions that do far more harm than good. We have now cataloged a long list of medical conditions that are recognized as autoimmune, and the list is still growing. Even poison ivy rash is a temporary autoimmune response against our skin cells (when altered by the chemicals from the plant). A more perfect immune system would surely not give rise in such a common and diverse list of "friendly fire" casualties and results.

Today, the greatest evidence of imperfection in organisms seems to reside within their genomes. As touched on elsewhere in this book (Chapters 5, 6, and 8), a significant fraction of the human genome does not seem to be concerned with the formation and function of the body. A clear and complete understanding of this question is still in the future, but we do know now that much of the genome consists of transposable elements that are generally considered to be (at least initially) genetic parasites, of pseudogenes of one type or another that contribute little if anything to the organism, and of DNA derived from retroviruses that incorporated their DNA into our ancient ancestors long before humans existed. Much study will be required to clarify our understanding of the large and diverse assemblage of non-gene DNA, but many of the workers in this area (see the John Avise quote at the start of this chapter) agree that the human genome is a huge collection of DNA from various sources, most of which originated as parasites, useless hitchhikers, or stowaways that contribute little if anything to the day-to-day utilitarian functioning of the genome. Certainly some of this non-coding DNA may rarely and by chance contribute something to the species (it can contribute to mutations—a few of which may be beneficial), but much of it appears to be useless and probably just along for the ride. The genome as currently understood clearly does not bear much resemblance to a perfected and well-tuned genetic unit.

In short, life seems to be awash in imperfection from the molecular to the organismal level. Even ecosystems are not the harmonious assemblages we once envisioned them to be (Chapter 19). With a nearly 4,000,000,000 year history, with change arising from accidental mistakes in molecular copying, with no overall plan, with evolutionary history obviously affected by uncountable contingent events—some of them disastrous cataclysms, with a great many life forms adapted to parasitize and damage other organisms, and with DNA itself ultimately concerned only with its own "selfish" survival, it is no wonder that life at all scales is shot through with imperfection. It was only because of ignorance, bias, and an incomplete view of reality that we ever thought otherwise.

Chapter 17

The Fossil Record and the History of Life

The fossil record is an amazing testimony to the power of evolution, with documentation of evolutionary transitions that Darwin could have only dreamed about.

Donald R. Prothero

That the sequence of fossils in the Earth's strata documents evolution is now accepted by scientists as an irrefutable truth.

Ernst Mayr

Some workers have said that we now have enough solid evidence that evolution has occurred even without the abundant visible evidence found in the fossil record. Still, it is convenient that the fossil record exists and that it so well supports the evolution over time of many groups of organisms. Only within my brief lifetime, an impressive array of important fossil finds have been made that have greatly advanced our understanding of the evolution and origin of humans, whales, the first land vertebrates, frogs, mammals, birds, and several other groups such as the plants. There is no doubt that a great many more important fossils will be found in the future—adding yet more important details to the fossil record and to our understanding of the history of life on our planet.

I will not in this brief chapter review the process of fossilization and its many variations, nor the aging of rock strata (determinations of age), nor the many other related topics that can be easily found in assorted other books and

Evolution. http://dx.doi.org/10.1016/B978-0-12-800348-0.00017-1
Copyright © 2014 Elsevier Inc. All rights reserved.

sources. Rather I will review briefly some of the important fossil finds that have filled important gaps in our understanding of the past history of life on Earth.

Like the increasingly rich fossil record for animals, significant plant fossils have been collected and analyzed, as well as good microfossils of prokaryotes and protists, but being a zoologist myself I mainly cover some of the significant finds in the animal record. In accord with our understanding of the pathway of evolution, the oldest clear fossils of organisms at more than three billion years old are, as expected, those of prokaryotes, many of which look almost identical to certain species of modern-day cyanobacteria. The next distinctly different organisms to show up at around two billion years old are the protists, with the first fossils of multicellular plants, fungi, and animals coming in later (again as expected) from about 0.7–0.6 billion years of age. In short, the broad record of "first appearances" fits precisely with the suspected path of evolution along the older main branches of the great tree of life.

The fossil record of animals was recently pushed back to about 6,500,000 years when some Australian fossils were analyzed and interpreted as sponges. Genomic studies, ultrastructure, and molecular clock data had also suggested that sponges were near the very base of the animal tree of life (Nielsen, 2012). A surviving group of protists known as the choanoflagellates are the likely sister group of animals, and their cells (colonial in some species) closely resemble certain cells of modern-day sponges.

At about 6,000,000 years old we have the Ediacaran fauna, also first discovered in Australia, with numerous macroscopic forms that were first believed to be animals of one type or another. The controversy continues as to exactly what they were—animals, algae, fungi, or some life form unlike anything surviving today. Whether some of these strange forms were animals or not, animals in at least the form of sponges, small worms, and other soft-bodied forms certainly existed by this time.

Coming up to either the famed Chengjiang fauna (around 520 million years old) or the Burgess Shale fauna (around 510 million years old), most of the modern phyla of animals appear in some diversity—all were marine creatures as no land forms had yet evolved. Among these strange creatures are the first arthropods, including the well-known trilobites, early crustaceans, and the strange *Anomalocaris* forms (Figure 17.1). A diversity of worms of one type or another were present, along with sponges, assorted cnidarians, early mollusks, and even a few small early chordates (the group that now contains all the vertebrates—including humans)—some in the form of a few pre-fish forms like *Myllokunmingia* (Figure 17.2). Because of these and other finds, we now know significantly more about the early beginnings of the animal kingdom that we did only 50 years ago, and new discoveries in this area continue to expand our understanding of early animal life.

Our knowledge of that most famed of all extinct groups, the dinosaurs, has grown significantly over the last 100 years, with most known species of dinosaurs having been discovered and described within that brief time. We now know of about 1,000 described species of dinosaurs, and the work continues to uncover more diversity and details of this fascinating group. Of course, the big news was that not

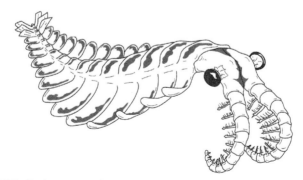

FIGURE 17.1 *Anomalocaris.* A strange early genus of arthropods from the Chengjiang fauna of China—not currently believed to be ancestral to any living arthropod group. Dated at about 525,000,000 years old. *(Figure by Jeff Dixon.)*

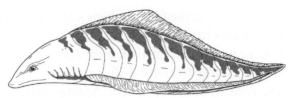

FIGURE 17.2 *Myllokunminga.* The earliest known fishlike chordate from the Chengjiang fauna of China. Dated at about 525,000,000 years old. Currently believed to be at least a close cousin of the ancestor that gave rise to all vertebrates. *(Figure by Jeff Dixon.)*

all dinosaurs went extinct. At least one line of the theropod dinosaurs evolved into birds, a group that includes more diversity than the dinosaurs likely ever achieved (today there are about 10,000 species of birds). Until the late 1990s, there were still a few holdouts on this conclusion who held that birds had evolved from an earlier "stem group" of reptiles that later gave rise to the dinosaurs. The evidence now is nearly conclusive that the later theropod dinosaurs were where birds originated, with fossils of many of the smaller bipedal theropods like *Sinornithosaurus*, *Caudiperyx*, *Microraptor*, and others showing clear impressions of simple-to-complex feathers covering much of their bodies—especially on their appendages, heads, and tails (Prothero, 2007). Also, many other traits once considered exclusive to birds, such as a furcula (wishbone) and hollow bones, have now been shown in several of the theropod dinosaurs. It is now suspected that even *Tyrannosaurus rex* may have had a few feathers—at least as youngsters. The new abundance of detailed fossil evidence in this area has made it very hard to draw a distinct demarcation line between dinosaurs and birds—exactly what is predicted when we have an abundance of fossil evidence about a transition from one kind of animal to another.

Another transition of interest that has become recently well documented is that of fish to amphibian, or fish to tetrapod—in short, aquatic to terrestrial vertebrates. Several other animal groups (arthropods, velvet worms, earthworms, land snails, etc.) independently adapted to terrestrial living—most much earlier than the vertebrates, but the fossil record for the vertebrate transition, which occurred about 365 million years ago, has recently become especially rich in transitional forms. For a long time we had known of fossil fish like *Eusthenopteron* and *Panderichthys*, which seemed ideal candidates for the early stages of this transition, and we knew of later amphibians like *Eryops*, which lived after the transition. Only in the last three decades or so have the likely intermediates in this story been found and recognized for what they are.

In order of limb transition from fin to limb, we now have good fossils of *Tiktaalik*, *Acanthostega*, *Icthyostega*, and *Hynerpeton* to bridge the gap between fish like *Panderichthys* and the later amphibians like *Eryops*, and there are an equal number of other genera that collectively have filled this former gap in the fossil record remarkably well (Prothero, 2007). Because of the possibility of several more yet undiscovered sister species, we will never be certain that any of the animals listed above were directly on the ancestral line leading to modern amphibians, and thus on to the reptiles, mammals, and birds. Nevertheless, all the evidence suggests that these animals were at least close relatives or "cousin species" to the actual ancestors of modern amphibians, and in turn of all land vertebrates.

The evolution of reptiles from amphibians is now well documented, as is the transition from reptiles to early mammals. There are in fact so many transitional fossils known for these two events that it is again hard to draw a line and say which of these animals was "the first reptile" or "the first mammal". We now also have several good transitional fossils showing how whales evolved from land animals, how frogs evolved from salamanders, how turtles evolved from nonturtle ancestors, and of course—how humans evolved from apelike ancestors. In the case of human evolution, the number of species now known from the fossil record has exploded within the past 50 years, making for a rather complex and bushy clade of species—all arrayed in time from smaller and small-brained apelike ancestors to larger and larger-brained humans. With new human finds still occurring every year or two, it will most likely be some time yet before a clear consensus forms as to our probable ancestry. It is interesting and instructive to ponder the idea that only 1–1.5 million years ago, there were probably three to five distinct human species alive at the same time (Prothero, 2007). In short, our relatively solitary position as the only living species in the human clade is a very recent occurrence.

I end this chapter by pointing out that the fossil record serves as a great example of the cumulative nature of science—meaning that, like science as a whole, as more and more data are collected and analyzed, our objective understanding of the world gains in completeness and accuracy. Science is somewhat like working on a huge puzzle for which you do not have a picture of the finished product. Also, a great many of the pieces of the puzzle are still undiscovered (as with the fossil record). As we find more of those pieces, and then identify where and how they fit into the puzzle, our view and understanding of the world and universe becomes more complete. This is the cumulative process in action which leads in a stepwise manner to a fuller understanding of our world—in the case of fossils, to a fuller understanding of evolutionary history.

Chapter 18

Contingency and Evolution

Contingency! This is the word that applies precisely to the whole of the evolutionary process.

Alexandre Meinesz

The course of evolution is only the summation of its fortuitous contingencies, not a pathway with predictable directions.

Stephen J. Gould

Many who reject the idea of evolution do so in part based on their misconception that evolution is a process based on chance occurrences alone. Since natural selection (Chapter 2) is not a chance process, these people are basing their rejection of evolution on a false premise. On the other hand, many who accept evolution as a real phenomenon in the living world have their own misconception that evolution is a progressive process that innately leads from simple "lower" forms of life to complex "higher" forms of life. This progressionist view of evolution was held by Lamarck and others of his day, who believed that all evolution was toward "higher" and more perfect forms. His explanation for the continued presence of a great many simple organisms was that they were constantly being spontaneously generated on the planet—to then start their independent evolution toward higher forms. Though few if any today accept the discredited notion of ongoing spontaneous generation, a great many otherwise educated people do still hold a progressionist view of evolution—a view that Darwinism effectively destroyed.

Almost all the adapted features of organisms are the result of natural selection, which, as already mentioned, is

Evolution. http://dx.doi.org/10.1016/B978-0-12-800348-0.00018-3
Copyright © 2014 Elsevier Inc. All rights reserved.

not a chance process. If natural selection were a chance process, then all variations in organisms would have an equal chance of being passed on into offspring, and I trust that by now it is clear that this is not the case. Natural selection is not a lottery where each ticket has an equal chance at the winnings. The lottery of life is stacked in favor of those individuals who happen to have inherited a successful genome, and not all individuals do so.

As opposed to being an innate process leading purposefully to particular features, traits, and abilities, evolution mainly tinkers with and continually shuffles the available genetic variation present within a species. Humans cannot quickly evolve wings sprouting angel-style from their backs because there is no genetic or morphological basis for the origin of such complex features. In those organisms that evolved wings, the wings always evolved from structures that were already present—but were originally not wings. This process is referred to as exaptation (discussed in Chapter 3). The answer for insects is still being conjectured, with wings having either evolved from segmental gill plate structures that originally functioned in aquatic larvae, or from thoracic overhangs of the cuticle that may have expanded first into a gliding foil, then later into functional wings, or from yet other segmental flap like structures (Daly et al. 1998). Whichever of these explanations turns out to be correct, insect wings almost certainly evolved from previous structures of some kind.

In vertebrates, of course, the wings of bats, birds, and pterosaurs all evolved from the already-present forelimbs that previously had served for terrestrial locomotion. Going further back, the four tetrapod limbs evolved from the four lateral fins of lobe-finned fish—the pectorals and the pelvics. The origin of lateral fins is more of a mystery because this origin happened a very long while back in evolutionary history, but even the earliest chordates in what appears to be the fish line (organisms like *Myllokunmingia* from the Chengjiang fauna in China) had lateral ventral ridge-like fins running most of the body length (Figure 17.2).

Perhaps these ridge fins simply broke up into separate fin units, two pairs of which were eventually retained in the majority of later fish.

All these example exaptations were therefore contingent on the availability of structures that could be "tinkered with" by mutations and genetic shuffling. Contingency means that the outcome of a process is dependent on previous conditions, starting points, available materials, etc. In his famed book *Wonderful Life* (1989), Stephen Jay Gould contrasted the American film classic "It's a Wonderful Life" (regularly shown several times every Christmas season) to the role of contingency in the history of life. In the film, the "hero" George Bailey discovers how his life had had major effects on the whole town of Bedford Falls. In other words, the town's outcome had been very much contingent on the life and activities of good old George. Of course, in true Hollywood style the effects of his life were more than a bit exaggerated to make for a more compelling story and film. However, few who have seem this film are likely come away with the notion that George Bailey's life was in any way destined to unfold the way it did—it just happened to occur as it did—and so followed the many contingent effects on the community.

In one part of Gould's book, he suggested that since the whole phylum Chordata, which includes all fish, amphibians, reptiles, and mammals (including humans), might easily never have survived and diversified, since early in the history of animal life only a very few small "insignificant" chordates were present along with a weird menagerie of other marine creatures. At least some of these creatures died out leaving no living descendants, and Gould argued that it might just as easily have been the early chordates that died out as opposed to some of the other unique animals, which, had they lived, might have given rise to unimagined phyla of diverse animal types that might have resembled nothing in the world we know today. There is almost certainly some truth in his point, though we will never know exactly why the chordate line survived and other lines perished in those far-distant Precambrian times.

In any case, life today is the result of literally millions of contingent events involving extinctions, geologic events, climatic changes, at least semi-random distributions and redistributions of life forms, mutations, horizontal gene transfers, genetic drift, symbiotic events, etc. Were a few land iguanas of South America destined to arrive on the Galapagos Islands and so give rise to the modern marine and land iguanas of that island group—or was that just a chance and contingent event? It was almost certainly the latter. The "seeding" of new islands with life forms is an especially clear example of contingency. In terms of species, which propagules of bacteria, protists, fungi, or lichens will arrive and prosper first? Which spores or seeds of plants will arrive on the island and in what order? Which birds, insects, rotifers, etc. will arrive and in what order? Anyone who understands even a little of ecology knows that the ecosystems present on islands today are the contingent results of what happened to arrive, when it arrived, and in what relative order those various species arrived.

Even over periods of only a few years, detailed prediction (one of the major goals in science) is nearly impossible for ecologists in terms of predicting the characteristics of a particular ecosystem say 3–5 years into the future—even in the absence of any human influences. Relative sizes of populations, relative species make-up, which diseases will be most prevalent, relative annual primary productivity, and several other ecosystem parameters are nearly impossible to predict more than a year or two out into the future. These parameters can be affected even over the short term by countless aspects of contingency such as weather, species movements—expansions and reductions, possible invasive species events, microbe and viral evolution, epidemics, and many other factors.

An analogy to weather (just mentioned) might be worthwhile. Meteorology is one of the true sciences, but it stands out among the sciences as being the one discipline where prediction, and especially long-term prediction, is difficult at best. Tomorrow's weather can be predicted with a fair level of certainty, as can the day after tomorrow. Though some predictions regularly go up to a week out, the weather even 3–4 days out from the prediction is often different from that predicted. The main reason for this is that there is a lot of contingency inherent in the complex air movements and temperature changes occurring in the atmosphere, and even with weather satellites, data collection for weather prediction is more limited than that required for accurate predictions in the other sciences.

Was the extinction of the dinosaurs by (at least in part) a huge asteroid striking the Earth 65,000,000 years ago a chance event, which had drastic contingent effects on the future evolution of land animals? Most certainly it was. From the level of random mutations in genes, to random duplication of genes, to random movements and displacements of individuals, to random asteroids striking the Earth, to random events affecting your own life—contingency is everywhere and ever-present. This ever-present aspect of contingency would be more obvious to us all if we could rid ourselves of the notion that we humans were destined to emerge from the evolutionary process. Nowhere in science is that notion supported. If this were so, then one might just as easily argue that each species present today was likewise destined to appear, and each with its specific morphology and behavior—a truly unlikely proposition.

The evolution of the bird fluke *Leucochloridium paradoxum* that manipulates the behavior of its intermediate host (a land snail) such that it exposes itself and its tentacles (antennae) to insectivorous birds (with at least one of its tentacles by this time swollen with a huge larval stage of the fluke—moving rhythmically to entice a bird to prey on the fluke larva) was almost certainly not destined to evolve, but instead is the result of a great many unknown contingent events in its evolutionary history. The same goes for each and every species on the planet.

A huge and contingent factor in the history of life on this planet has been that of continental drift. The convection currents in the Earth's hot molten interior that drive the movements of the crustal plates, though they can remain consistent for long spans of time, do nonetheless change unpredictably over geologic time. In one era two plates are driven together, in another they separate or break at new boundaries to move in other directions. As they do so, thousands of terrestrial species are contingently separated by the spreading, or brought together in sympatry during collisions. These events typically lead to speciations, extinctions, and new selective pressures of all kinds that have significant effects on the direction of evolution in the affected species. The direction of movement can also move a continent into a new climatic zone. Antarctica was once located closer to the equator, and today fossils of reptiles and ferns can be found in the rock layers of that continent—having evolved there prior to its contingent southerly sojourn.

One of the most thought-provoking points concerning contingency is the contingency of our own appearance and existence on the Earth. Those with religious and other metaphysical beliefs largely hold the notion that humans (or something approximating humans in body form and intelligence) were a probable-to-certain outcome of the evolutionary process—on this, or any other planet where life gets a foothold. Most science fiction novels and movies are top-heavy with this unfounded assumption, with the many incarnations of the Star Trek series being perhaps the worst offender.

Even some in the scientific community hold to the idea that intelligent humanoid-like organisms would be sure to arise eventually in the evolution of life on Earth or on other planets (Morris, 2005). Since humans are but one species among many millions that have existed on this planet, to say that the evolution of our species was destiny is again basically the same thing as saying that the beef tapeworm, the white egret orchid, the death angel mushroom, or the pygmy seahorse were all destined to appear—given enough time. As soon as we switch the emphasis from our species to any one of the other millions of living species, the absurdity of this whole idea should become apparent to anyone who considers the contingency involved in the long and complex history of evolution on this planet—along with the fact that natural selection has *only* the simple effect of adapting organisms to survive and reproduce effectively in their local environment. There are obviously many millions of unique ways to achieve those simple ends. The human way is but one of the many millions of contingency-based solutions for achieving those ends. Along with several others discussed in this book, contingency plainly must be recognized as one of the major factors that explains the resulting particulars of the evolution of life on Earth.

Chapter 19

Opportunity

If natural selection is the main factor that brings evolution about, any number of species is understandable: natural selection does not work according to a foreordained plan, and species are produced not because they are needed for some purpose but simply because there is an environmental opportunity and genetic wherewithal to make them possible.

Theodosius Dobzhansky

In effect mass extinctions provide opportunities for the survivors to radiate into the vacated niches left behind by the extinct species.

Paul B. Wignall

Earlier progressionist thinking not only held that evolution was innately about progression from lower to higher forms but also held that species filled set roles in natural ecosystems, with each playing its necessary part in a harmonious whole. The lower-to-higher thinking was effectively made unscientific by Darwin's ideas, but it took much longer for biologists to shake the idea of ecological roles for each species. Certainly you can recognize a tiger as a top-level predator, and tigers may even have some desirable effects (again in our minds) on the jungles where, sadly, too few still survive. Plants obviously produce oxygen that many other organisms (including plants) require. They also often serve as food for a great many other organisms. So, from examples like these we are easily led to believe that tigers and plants and many other species have particular roles to play, and evolution might be all about ensuring that species evolve to fill those "essential" roles.

Evolution. http://dx.doi.org/10.1016/B978-0-12-800348-0.00019-5
Copyright © 2014 Elsevier Inc. All rights reserved.

Once again though, we have stepped outside of scientific thinking in assuming that evolution is working to create ecosystems in which each species fills some required role, because that would assume some plan or order was being "fulfilled"—and where would such a plan originate? This sounds too much like teleology, destiny, and metaphysics—decidedly unscientific concepts to say the least. Evolution in any species or population is primarily about improved survival and reproduction—ultimately the passing of copies of one's genes into the future. Many species are known to achieve these objectives without filling any obvious or imaginable role whatsoever in their environment. One of the three conventional subclasses of symbiosis is commensalism. A commensal is a species that needs another species for its survival, but it neither harms nor aids that other species in any way. There are many known (and undoubtedly a far larger number of unknown) amoebae and flagellates that live as commensals in the intestines of animals. Many of these cannot survive actively in the outer environment, and exist there only as dormant and resistant microscopic cysts that lie or drift in the environment for a time on the off-chance that another potential animal host will accidentally ingest that cyst. If this occurs, the dormant organism will break out of its cyst and become active in the new host intestine and go back to its business of surviving on a few gut bacteria or sloughed host cells—again doing no measurable harm, and providing no benefit to its host worm, fish, rat, human, etc.

If you consider such species for only a moment, it becomes clear that they almost certainly are not filling any important role in nature—having essentially no effect at all on the ecosystem of which they are a part. Such commensals exist because they found an "opening" or an opportunity in the community of living creatures, and they evolved to take advantage of (adapted to) that opening. There is no evidence for any force nudging species to evolve into commensals of this type, but there are obviously many opportunities for them to do so. If you still feel that surely everything must have a role in nature, then ask yourself what role the two-thirds of known species that are classified as parasites are filling. Parasitism is the dominant lifestyle on planet Earth, with every known non-parasitic species being plagued by several parasitic species—many of which are known to be endemic to only one host species. More telling still, from a scientific viewpoint, ask yourself what essential role humans are playing in the world. It is true that a million years ago it may have been conceivable to view humans as filling some loosely defined role in nature, but today our only dominate effect on the environment is one of destruction on a massive scale. So, is that our "role" in nature—to destroy it? I don't think so.

If one compares somewhat similar ecosystems in widely spaced locations on Earth, one would find that for those organisms we can define as pollinators, top-level carnivores, dominant producers, etc.—these would not be the same exact species in Australia as they are in South America, or as they are in Africa, etc. Why should each ecosystem be filled with species that are largely different from those that occur elsewhere in the world? If there were "a plan" for filling set roles in ecosystems, why wouldn't similar roles across ecosystems all be filled with the same species? Darwin himself asked and pondered these questions. Simply put, the answer is that in evolution there is no plan—only opportunities, yet sometimes very similar opportunities in two or more different locations.

After a number of free-living species are present in the environment, the opportunity for mutualists, commensals, and parasites opens up. After several species adapt to be "plant-eaters", the opportunity for carnivores opens up, and carnivores will then evolve to take advantage of that opportunity. After ticks and fleas evolve to specialize on large grazing mammals, the opportunity for oxpeckers or similar birds that make a living feeding on these ectoparasites opens up. After an oceanic island forms owing to volcanic activity, all manner of opportunities exists for its colonization by a host of spores, seeds, birds, insects, etc. And so it goes—a never-ending series of newly occurring opportunities that accompany the many changes that occur in the various environments of the planet over geologic and even shorter time spans. Though we usually decry the damage done by invasive species such as rats on the Pacific islands or kudzu in the United States, these are examples of new opportunities for the rats and the kudzu vine when they are introduced into habitats where they can compete with, and even outcompete, the native species.

Another argument in support of the importance of opportunity is that whenever opportunity exists for life forms to move in, they seem to do so. When the last great extinction event wiped out the last of the dinosaurs 65,000,000 years ago, many opportunities were opened up in the post-dinosaur world, which the surviving mammal lines quickly started to diversify and adapt into. This same event also wiped out the several large, predacious marine reptiles like the mosasaurs, the plesiosaurs, and the ichthyosaurs. This vacancy is hypothesized to have been the opening that "allowed" the evolution of whales from land mammals to take place in a period of only the last 50,000,000 years—to take advantage of that set of opportunities—and tellingly all the first whales were predators—just like the reptiles they replaced. Another similar example might be the increased diversification of marine crustaceans following the extinction of the last trilobites 250,000,000 years ago in the great Permian extinction event.

Returning to the realm of parasitism, recall again that most of the species on Earth are parasites. Why so many kinds of parasites? The most obvious and logical answer is opportunity. If organisms can be parasitized (and they all seemingly can), then every species provides an opportunity for parasitism to evolve. Even parasites can have parasites. Whether you consider viruses to be alive or not, they do evolve and diversify to specialize in different host species. In writing about the diversity of marine viruses, Carl Zimmer stated: "One reason for all this diversity is that marine viruses have so many hosts to infect" (Zimmer, 2011). In Chapter 13, the case of the hawthorn fruit fly was explained in that the introduction of apple trees provided the opportunity to parasitize (fruit flies can be said to parasitize the fruit of these trees) a new resource—the apple tree, thus setting in motion a case of sympatric speciation.

Even bacteria and archaea are parasitized by viruses. Plants are parasitized by other plants, by insects, by fungi, etc. We do not know much yet about the parasites of fungi, but undoubtedly they are plagued by a diversity of parasites as well. Animals are the best known kingdom in terms of their parasites. Only plants fail to parasitize animals. Humans along with all other animals are parasitized by numerous viruses, bacteria, protists, fungi, and other animals. Parasitism is, of course, not a taxonomic grouping of related forms. Parasitism is a lifestyle that has evolved independently countless times throughout the history of life in many different kinds of organisms—all due to the many available opportunities to live as a parasite. Even some vertebrates have taken up the parasitic lifestyle, including some catfish (candiru), some sharks (cookie-cutter sharks), some bats (vampire bats), some birds (brood parasitism in cuckoos), and others.

At the level of the genome, whenever one or more genes are duplicated by chromosomal mutations, an opportunity is then available for one of the two copies to mutate and create a new protein or a new regulatory switch that just might be advantageous to the affected organism/species, and perhaps result in altered phenotypes. Though most such mutational events most likely created useless pseudogenes, some have turned out to be useful in generating more complexity and more biodiversity in the living world—in short, more opportunities seized.

Biodiversity itself then is a major source of these opportunities. Hosts from all kingdoms provide the opportunity for parasites to evolve. The world's trees have provided the opportunity for the evolution of epiphytes, of climbing vines, of arboreal monkeys and insects, of shade-tolerant species that grow beneath them, etc. Coral reefs provide countless opportunities for species to adapt to live in some association with the physical and biological structure that is a reef. Many species feed on the corals, some hide in the coral crevices, some forage and hunt there, some use the reef as a spawning area, and many kinds of sessile organisms use the solid reef as an attachment base on which to grow. To quote one of the greats in evolutionary biology: "Species are produced not because they are needed for some purpose but simply because there is an environmental opportunity and genetic wherewithal to make them possible" (Dobzhansky, 1973).

Humans certainly took advantage of opportunities as they opened up to expand their range and develop new survival aids, many of which were only "seen" because of the expanding mental landscape of our earlier ancestors: the opportunity to control and create fire; the opportunity to create useful tools from rocks, bones, and antlers; the opportunity to keep warm by fashioning clothing from animal skins; the opportunity to develop agriculture and permanent settlements; the opportunity to trade with other groups; the opportunity to move over the water in rafts and boats—the list goes on and on. These more recent developments of our cultural evolution seem to be the major reasons behind our successful spread and domination of the planet. We still search mentally for opportunities in the realms of jobs, education, technology, industry, marketing/trade, fashion, sports, grants/scholarships, etc.—always looking for an opening that others have yet to fill. We, like all other species, are here, and we are what we are in large part because of countless opportunities that were seized over our long evolutionary history.

Chapter 20

Phylogeny—The Tree of Life

Our classifications will come to be, as far as they can be made so, genealogies.

Charles Darwin

The TOL (tree of life) has become the centerpiece of evolutionary biology and, in a sense, of biology in general.

Eugene V. Koonin

Surely one of the most amazing things to consider about the accepted evolutionary view, is that all of Earth's species are literally genetic cousins. Some are very close cousins, like chimpanzees, bonobos, and humans. Some are more distant cousins like ferns, horsetails, and conifers, and some are extremely distant cousins like bacteria, yeast, and bluebirds. This understanding of common descent gives biologists the potential to discover and understand the actual pattern of ancestry among the world's species, such that a family tree or "tree of life" can be elucidated. At least that is the hope, and we now have seemingly all the basic tools necessary to examine every bit of available evidence needed to work out these ancestral relationships—the phylogeny of life. Accumulating all of the actual genomic and other data needed for this task will take us into the foreseeable future.

Ernst Hackel, the famed German biologist and artist, made some of the first attempts to conceive and illustrate this tree of relationships for Earth's diverse species. Considering that he was working in the late 1800s, he did a remarkably reasonable job of positioning recognized organism groupings into his trees of relationships. Of course, this

Evolution. http://dx.doi.org/10.1016/B978-0-12-800348-0.00020-1
Copyright © 2014 Elsevier Inc. All rights reserved.

was before the fossil record was well known, before development or ultrastructure could be adequately studied, before the age of genomics, and before we had a good method for deciding what traits and organisms were to be considered old or ancestral, or newly derived.

The tree of life project is today much more complete, and is fleshed out with the huge array of diverse organisms now known to exist. A modern tree is certainly far more complete and accurate than Hackel's early trees, though there is still much more work to be done through studies of development, ultrastructure, and genomics. The fossil record continues to be important as new fossil finds add support to existing hypotheses, or correct some mistaken notions about relationships. Also, at least two-thirds of the species on Earth are believed to be as yet undiscovered, with some of these unknowns possibly constituting higher-level taxa. This was the case with the animal phylum Cycliophora, which was first discovered only in 1995. Understanding these myriad relationships is a huge undertaking that biologists believe to be of great importance. Only by discovering the actual phylogenetic relationships of all species will we ever understand the path of life's evolution on Earth—the history of Earthly life.

The word phylogeny can be defined as the evolutionary history of a group of related species. Since all species are related, the complete tree of life is the largest scale phylogeny we can represent. Most illustrated phylogenies in the literature concern smaller related groups or clades like the Animal Kingdom, the vertebrates, the amniotes, the snakes, or the rattlesnakes. Each of the groups just mentioned is obviously nested within the preceding group, with the Animal Kingdom being the most inclusive clade, but an illustration of phylogeny could be done at any of these levels or scales. The phylogeny of the rattlesnakes would go right down to the species level, while most illustrated phylogenies of the Animal Kingdom would stop at the phylum or class level. Such illustrations are referred to as cladograms, because they attempt to illustrate the historical cladogenesis or splitting of the subgroups shown. Figure 20.1 shows a fairly complete cladogram for the amniotes, the clade that includes the mammals, birds, and several of the "reptile" groups.

How do scientists go about constructing these clado-grams? How do they know where the various groups should be placed within the cladogram? They do so based on the comparison of a number of traits or characteristics that can be identified and compared between the several groups to be represented in the cladogram. These characteristics can be morphological, physiological, metabolic, behavioral, or molecular. Cladograms once relied primarily on morphology, and those for extinct groups (present now only as fossils) must largely still do so, but for recent and living groups, ultrastructure (the microscopic structure of cells and tissues) plus direct protein and DNA comparisons have become the standard, with DNA itself being the most character-rich source of phylogenetic information.

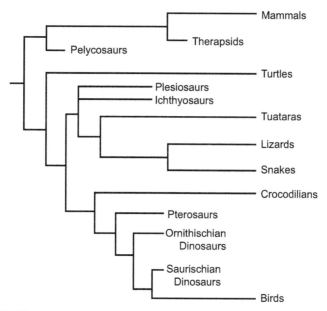

FIGURE 20.1 A phylogenetic cladogram showing most of the major amniote vertebrate groups and their relations to one another. *(Compiled from several sources.)*

Actually it is always preferable to include as many characteristics as possible—a mix of molecular, morphological, physiological, etc. In addition to comparing characteristics between and among groups, the characteristics need to be identified as either old ancestral characters, or newly derived characters. This concept is referred to as the polarity of the trait—polarity in time, that is.

Polarity is based on the simple fact that some of the characteristics or traits of organisms are very old in evolutionary terms, while some are very new. Deciding which characters are old and which are new and then measuring the general ratio between old and new characters in a species or higher grouping are essential steps in discovering its relationship to other species or groups, both those closely related and those more distantly related.

Several sets of terms have been used since Darwin's time to refer to characters that can be compared as old versus new. We need not review all of these, since some are rarely used today. Perhaps the two least desirable sets of terms for this concept were primitive and advanced—and lower and higher. Primitive and lower were the designation for older traits in a line of descent, while advanced and higher referred to newer and thus younger traits or characters. The problem with these terms should be self-evident. They seem to imply value judgments in that the primitive or lower characters sound like they are not as good or successful as the "higher" or "advanced" characters. Modern biologists generally try to avoid using any terms that imply that some species or its traits are better or worse than alternative traits in other species, though unfortunately a few continue to do so. Traits may indeed be old ones, but if they are still present in living organisms, they are obviously still working and serving their role well.

Two preferable sets of terms for this same concept would be "ancestral" and "derived"—and plesiomorphic and apomorphic. Here ancestral and plesiomorphic refer to older traits, while derived and apomorphic refer to newer or younger traits. Plesiomorphic and apomorphic are obviously the more technical terms used almost exclusively by biologists in their scientific papers and meetings. Ancestral and derived have the advantage of being relatively free of implied value judgments, and are certainly easier to remember than plesiomorphic and apomorphic.

What are some examples of polarity? In the plants, reproducing asexually or by spores or both is the older trait, still found in a number of plant groups like mosses and ferns. Some plants later evolved seeds—a newer, derived reproductive method. So within plants, the seed-producing plants are considered derived compared to the ancestral plants that still use spores. Another similar example would be the absence of wings in some insects like silverfish—versus the presence of wings in most other insect groups. The first insects did not have wings, and silverfish are thought to have evolved from a branch of these early insects without ever having evolved wings. Most modern insects, however, are descended from another early ancestral line in which wings did later evolve—which proved to be a very successful adaptation that gave rise to an incredible number of descendent species of winged insects. Another example mentioned elsewhere in this book are the rattles of rattlesnakes. Most snakes lack rattles, and surely the first snakes did not have them. Rattles are a newly derived structure found only in the rattlesnakes, so the ancestral condition would be a lack of rattles, while the derived condition in rattlesnakes is the presence of rattles. This last example especially illustrates both of the old "rules of thumb" for determining whether a trait is ancestral or derived. In most cases, the ancestral trait will be common or shared by a great number of species, while the derived trait will be rare and found in only a few species. There are close to 3,000 species of snakes known, but only about 35 species of rattlesnakes—fitting nicely the rule of thumb that ancestral traits tend to be common within a clade, while derived traits tend to be restricted and rare. The other rule of thumb says that ancestral traits tend to be simple while derived traits tend to be more complex. Having rattles is certainly more complex than lacking rattles, so here again the rule of thumb holds when rattlesnakes are compared to other snakes. These are not hard and fast rules because there are exceptions to each, but still they are generally useful in phylogenetic studies.

This common/rare rule is also important in direct genomic comparisons. This will be a hypothetical example, but if say most snakes had the DNA base guanine (G) at a particular position in a relatively conserved gene, while only rattlesnakes had thymine (T) in the same homologous position, this would imply that thymine was a derived base in that position in the ancestor of the rattlesnake group or clade. In other words, a mutation had changed guanine to thymine in that gene in that ancestor with the result being that today's rattlesnakes all inherited that derived base. There exists a significant amount of more technical jargon for these concepts, which can be found in more technical works dealing systematics and phylogeny.

A great deal has been learned about the phylogeny of life within the past 30 years or so, and undoubtedly a great deal more will be learned in the near future. Knowledge of phylogeny does more than satisfy the curious mind. Phylogenies of discrete populations of HIV, anthrax bacteria, flu viruses, etc. can aid governments and medical groups in tracking and identifying sources of disease organisms, developing vaccines and other drugs, and even identifying terrorist sources of biological weapons. Phylogenies of modern crop plants can point the way back in time and aid in identifying the ancestral species of our various crop species and varieties—knowledge that can be useful in plant breeding and crop improvement. In short, elucidating accurate phylogenies is not just an academic pursuit. It has assorted practical applications as well.

Chapter 21

Progress — Purpose?

Evolution is not a race for progress with the goal of achieving perfect form and function. Rather it is a race for survival.

Alexandre Meinesz

Scientists have little time for subjective judgments. Evolutionary progress therefore is now not much discussed.

Mark Ridley

...how can we say a wasp is "higher" than a worm, or a man "higher" than either, or even than an amoeba, when all manage to exist and survive?

Julian Huxley

One of the oldest and most common, yet false, ideas concerning evolution is that it is an innately progressive process that will naturally lead to organisms that are more complex, more intelligent, and according to some—even more humanoid. Steven Weinberg wrote: "The more the universe seems comprehensible, the more it also seems pointless" (Weinberg, 1993). Many modern scientists would basically agree with this statement. The universe and life are indeed changing and evolving through time, but only a biased human mind could see the process or the results of this change as indicative of any type of progress or purpose. Progress is an entirely human concept that many humans have applied in this case to a process decidedly not concerned with what humans are—or what they think.

Additionally, there simply is no (and probably can be no) objectively based definition of the term progress as

Evolution. http://dx.doi.org/10.1016/B978-0-12-800348-0.00021-3
Copyright © 2014 Elsevier Inc. All rights reserved.

applied to the phenomenon of life. Rather, progress is a subjective notion that can be defined in countless ways, some of which might even be contradictory. The very concept of progress involves the idea that something is getting "better". As has been pointed out by many previous workers, science strives to describe and understand what is, not what should be. Any thinking directed at what "should be" involves concepts of good versus bad, or better versus worse. In short, this line of thinking involves a value judgment—something that science is unable to arrive at. Values are completely outside of the realm of science, except for the few basic values that scientists simply accept as guidelines for their work. Briefly the four core values science holds are:

- Curiosity is good
- Knowledge is good
- The fabrication or falsification of data is bad
- It is good to keep an open mind, tempered with a healthy dose of skepticism

These values obviously involve a judgment on what is good or bad, but they do not touch in any way on what is good or bad in the natural universe outside of the actual work and practice of science. You might believe the moon is beautiful, but this is not a scientific statement or conclusion. You might believe warm-blooded (endothermic) vertebrates are more "advanced" than cold-blooded (ectothermic) vertebrates, but this too is not really a statement based in science because of what seems to be implied by the word advanced. Some would argue that by their definition of advanced, mammals are more advanced than reptiles, but again the term dangerously implies a value judgment. Mammals and reptiles can certainly be compared in many respects that might involve the terms faster, more efficient, or more complex. These could be perfectly valid scientific statements involving description, because the wording need not imply that faster or more complex is in some way better than slower or less complex.

It is unfortunately not hard to find wording in even some biology textbooks that strongly implies some of the values associated with progressionist thinking (I know because I have caught such wording when reviewing textbooks), but however often such implications crop up, they are not science based and really should not be used or implied by science writers. If you state that mature longleaf pine trees are taller than mature dogwoods, you have made a simple statement of fact without implying that pine trees are in some way better than dogwoods. But when someone says that a wolf is more advanced than a fence lizard because it can control its body temperature more precisely, all they have really done is to state a single difference in the physiology of a mammal versus a reptile. If a biologist were asked whether a wolf was a better organism than a fence lizard, he should certainly respond that the question makes no sense—with the word "better" ringing an alarm of unscientific thinking. Differences alone do not imply a case of better versus worse (racists take note!).

Many biologists slip into the habit of using the phrases more advanced, more efficient, etc., and some may actually believe that mammals are better in some qualities than reptiles, but again this is not scientific thinking at all. Rather these are value-laden ideas and language that reveal some bias in the writer. To quote Lynn Margulis: "All beings alive today are equally evolved. All have survived over three thousand million years of evolution from common bacterial ancestors" (Margulis, 1998). Biologists should all know by now that the only goal of living organisms is to survive and pass their genes into the next generation. Bacteria are still doing that job extremely well, and they are little evolved morphologically from their ancestors of some 3,500,000,000 years ago.

To believe that life is striving toward goals of greater complexity, specific form, or superior intelligence is to partake of the progressionist philosophy of Lamarck and others of his day, who believed that evolution consisted of a series of changes that led to progressively "higher" forms. Darwinian thinking basically destroyed any foundations for such a belief, and surveying the diversity of life around us today likewise negates such thinking. The late Stephen J. Gould did an excellent job of explaining why progressionist thinking is false and unfounded in his excellent book *Full House* (Gould, 1996), and I would recommend that book to anyone interested in a deeper explanation of these ideas.

Most of the species that have ever lived have become extinct without giving rise to surviving lineages. Trilobites numbered more than 5,000 species and were successful for many millions of years, but the last trilobites died out 250,000,000 years ago in the great Permian mass extinction event, with none known to have left surviving lineages. Were trilobites in some way not as successful as the creatures that eventually replaced them (mostly crustaceans)? Had trilobites survived the Permian extinction event, would they necessarily have evolved into significantly "higher" forms? The answer to both these questions is—very doubtful. An excellent quote touching on this subject is one by George Gaylord Simpson: "There is no clear overall progression. Organisms diversify into literally millions of species, then the vast majority of those species perish and other millions take their places for an eon until they too are replaced. If that is a foreordained plan, it is an oddly ineffective one" (Simpson, 1964).

I would also refer you to Chapter 16 and its numerous examples of imperfections in living organisms; to Chapter 6, which covered those many parasitic transposable genetic elements (at least originally) that make up such significant fractions of eukaryotic genomes; to Chapter 8, which concerned the large role that neutral evolution has played in genomic evolution; and to Chapter 9, which attempted to show the large role of random genetic drift in evolution. In various other places in this book the significant role of viruses and other parasites on evolution has been mentioned. Collectively, all these sources of largely undirected (so far as anyone can see) evolutionary change certainly do not argue in favor of an underlying progress or purpose in the evolutionary process—aside from that immediate purpose and tune that all organisms march to: survival and the attainment of genetic fitness.

As was stated earlier, this lack of a plan, of a purpose, of set roles to be played, or of goals in the evolutionary process is precisely why life has been free to evolve into any form, lifestyle, size, metabolism, life history, etc. that "works" to promote successful survival and reproduction. There have obviously been many billions of adaptations to many millions of different ways of attaining fitness, or of staying in the game. It might seem that insects have certain advantages over the slower growing and slower reproducing elephants, but both insects and elephants are here, and both are obviously successful because of their very different but respectively successful sets of adaptations. This freedom from a plan or goal is arguably the central reason behind the incomprehensible vastness of biodiversity we find in the living world. Hopefully we will some day better recognize the riches of our living world and work to preserve its amazing biodiversity while continuing to learn of its awesome secrets—all without worrying about unwarranted questions of purpose or progress.

When Halley's Comet makes its return in 2061, astronomers will have at their disposal the most sophisticated instruments ever to be trained on this old friend—last seen in 1986. Undoubtedly we will learn far more about this regular visitor to our skies than ever before, and excitement will run high for those involved in the data collection. I would venture to say that no one involved at that time will be asking what the purpose of this comet is, or what role it plays in the big scheme of things. So why do so many continue to do so in the realm of living things? The only reasons I can venture are because we are living things, we have an inflated opinion of ourselves, and we like to believe our lives individually and collectively serve some grander purpose. However, believing and wishing something does not necessarily make it so, especially in the absence of any clear and unequivocal supporting evidence.

We humans, however, do have the wonderful and amazing ability to create our own meaningful plans and purposes, both individually (getting a college degree, planning that dream vacation) and collectively (a family reunion, a new school, a government). The ability to form long-range plans and work toward them is one of several traits (including complex language, writing, science) that set us apart from the other species of this planet. These abilities have obviously led to great accomplishments, but also to great evils (the Holocaust). Scientists believe it is good to increase our knowledge of the world and universe we inhabit. How we use that knowledge and apply it to our plans and purposes is entirely up to us. More often than not, ignoring or hiding real knowledge is a source of events and viewpoints that are often judged as negative or destructive—that belittle or harm humanity.

References

Andersson, J.O., 2006. Genome evolution of anaerobic protists: metabolic adaptation via gene acquisition. In: Katz, L.A., Bhattacharya, D. (Eds.), Genomics and Evolution of Microbial Eukaryotes. Oxford University Press, Oxford.

Avise, J.C., 2010. Inside the Human Genome. Oxford University Press, Oxford.

Bergstrom, C.T., Dugatkin, L.A., 2012. Evolution. W.W. Norton & Company, New York.

Burt, A., Trivers, R., 2006. Genes in Conflict. The Belknap Press of Harvard University Press, Cambridge.

Chandler, A.C., 1940. Introduction to Parasitology. John Wiley & Sons, Inc, New York.

Conrod, W., Rosenblum, E.B., 2008. A desert Galapaos. Nat. Hist. 117 (4).

Daly, H.V., Doyen, J.T., Purcell, A.H., 1998. Introduction to Insect Biology and Diversity. Oxford, Oxford University Press.

Danchin, E.G., 2011. What nematode genomes tell us about the importance of horizontal gene transfers in the evolutionary history of animals. Mobile Genetic Elements. 1:269–273; http://dx.doi.org/10.4161.

Dawkins, R., 1989. The Selfish Gene. Oxford University Press, Oxford.

Dawkins, R., 1998. Unweaving the Rainbow. Houghton Mifflin Company, Boston.

Dobzhansky, T., 1973. Nothing in biology makes sense except in the light of evolution. Am. Biol. Teach. 35 (3).

Fairbanks, D.J., 2007. Relics of Eden: The Powerful Evidence of Evolution in Human DNA. Prometheus Books, Amherst, NY.

Fontdevila, A., 2011. The Dynamic Genome: A Darwinian Approach. Oxford University Press, Oxford.

Futuyma, D.J., 2013. Evolution, third ed. Sinauer Associates, Inc, Sunderland MA.

Gould, S.J., 1994. This power of this view of life. Nat. Hist. 103 (6).

Gould, S.J., 1989. Wonderful Life: The Burgess Shale and the Nature of History. W.W. Norton & Company, New York.

Gould, S.J., 1996. Full House. Harmony Books, New York.

Graham, L.A., Loughee, S.C., Ewart, K.V., Davies, P.L., 2008. Lateral transfer of a lectin- like antifreeze protein gene in fishes. PLoS ONE 3 (7), e2616. 10.1371/journal.pone.0002616.

Huang, J., Kissinger, J.C., 2006. Horizontal and intracellular gene transfer in the Apicomplexa: the scope and functional consequences. In: Katz, L.A., Bhattacharya, D. (Eds.), Genomics and Evolution of Microbial Eukaryotes. Oxford University Press, Oxford.

Holldobler, B., Wilson, E.O., 2009. The Superorganism: The Beauty, Elegance, and Strangeness of Insect Societies. W.W. Norton & Company, New York.

Jablonka, E., Lamb, M.J., 2005. Evolution in Four Dimensions. The MIT Press, Cambridge.

Lane, N., 2005. Power, Sex, Suicide: Mitochondria and the Meaning of Life. Oxford University Press, Oxford.

Lecointre, G., Le Guyader, H., 2006. The Tree of Life: A Phylogenetic Classification. The Belknap Press, Cambridge.

Margulis, L., 1998. Symbiotic Planet. Basic Books, New York.

Mayr, E., 1997. This is Biology: The Science of the Living World. The Belknap Press, Cambridge.

Mock, D.W., Drummond, H., Stinson, C.H., 2010. Avian Siblicide. In: Sherman, P.W., Alcock, J. (Eds.), Exploring Animal Behavior. Sinauer Associates, Inc, Sunderland, MA.

Morris, S.C., 2005. Life's Solution: Inevitable Humans in a Lonely Universe. Cambridge University Press, Cambridge.

Nielsen, C., 2012. Animal Evolution: Interrelationships of the Living Phyla, third ed. Oxford University Press, Oxford.

Prothero, D.R., 2007. Evolution: What the Fossils Say and Why it Matters. Columbia University Press, New York.

Rudge, D.W., 2005. The beauty of Ketterwell's classic experimental demonstration of natural selection. Bioscience 55 (4).

Schilthuizen, M., 2001. Frogs, Flies, and Dandelions. Oxford University Press, Oxford.

Scott, P., 2008. Physiology and Behaviour of Plants. John Wiley & Sons, Ltd, Hoboken NJ.

Shapiro, J.A., 2011. Evolution: A View from the 21st Century. FT Press Science, Upper Saddle River, NJ.

Simpson, G.G., 1964. This View of Life. Harcourt, Brace, & World, Inc, New York.

Van Dover, C.L., 2000. The Ecology of Deep-Sea Hydrothermal Vents. Princeton University Press, Princeton.

Weinberg, S., 1993. The First Three Minutes (updated second ed.). Basic Books, New York.

Williams, G.C., 1996. Plan and Purpose in Nature. Weidenfeld & Nicolson, London.

Zimmer, C., 2005. Smithsonian Intimate Guide to Human Origins. Madison Press Books, Toronto.

Zimmer, C., 2011. A Planet of Viruses. The University of Chicago Press, Chicago.

Zimmer, C., Emlen, D.J., 2013. Evolution: Making Sense of Life. Roberts and Company, Greenwood Village, CO.

MOST OF THE SOURCES OF EVOLUTIONARY CHANGE—A SYNOPSIS

This abbreviated list includes most of the mechanisms and factors that contribute directly or indirectly to genomic variability and evolutionary change in populations.

1. **The Genetic Variations Present in Most Species:** Typically in the form of alleles. Most of the change involved in creating the large number of dog breeds out of the natural wolf population by artificial selection involved purifying respective combinations of variations already present in those natural wolf populations. Nature can bring about similar outcomes from variation already present in most species. (Chapter 5)

2. **Mutations:** (Both point and chromosomal) Mutations are the ultimate source of both new genes and new alleles which become the raw material of evolutionary change. Duplications of chromosome segments housing genes create new gene copies, some of which may evolve unique functions and so become new genes. Deletion mutations are responsible for gene loss which is likewise an important factor in evolution, especially so in certain kinds of organisms such as some obligate parasites. (Chapter 5)

3. **Polyploidy:** The ultimate chromosomal mutation in which the whole genome is duplicated one or more times. This rare occurrence has had major effects on the evolution of some eukaryotic groups, especially in the plants, but also in certain animal groups like the vertebrates. (Chapters 5 & 13)

4. **Natural Selection:** Darwin's greatest achievement— the first known credible mechanism of evolutionary change, and still the only one that can craft adaptations in species. Basically natural selection is a simple

idea that revolutionized biology in general (Chapter 2). Natural selection subsumes sexual selection which is a large but distinct subcategory of natural selection. Darwin was the first to recognize the importance of this important force in evolution.

5. Diploidy: In diploid organisms rare recessive alleles can exist even if harmful to some degree, especially in large populations where only a few individuals are carriers of these rare alleles. Under some changing environmental conditions, some of these rare alleles may prove beneficial. Without diploidy, these once harmful alleles would have been extinguished by natural selection. In short, diploidy allows more variation to exist in the gene pool.

6. Sex: Diploidy above is typically tied to meiosis somewhere in the life cycle, which shuffles genes and alleles extensively in the production of gametes (or spores)—resulting in vast potential for variation in the offspring of sexually reproducing species—one of the foundational points from which Darwin and Wallace derived the idea of natural selection.

7. Changing Environments: If environments were stable for millions of generations, evolution would largely come to rest with most traits existing under stabilizing selection for the adequate status quo. Only with environmental change can evolution produce its larger and more obvious effects. Environments have most definitely changed across the face of the Earth over geologic time. Environments also change for certain species when they are relocated into new and different environments, as when a few ancestral South American iguanas were washed up on the Galapagos islands some 3,000,000 years ago or so—entering directional selection in several traits leading to their adaptation to this very different environment. (Chapter 10)

8. Migration: When organisms of the same species move away from a population's range (emigration), they may affect the frequency of certain alleles or other genomic

variations remaining in that population. Likewise, if organisms move from other distant populations into a local population (immigration), they may bring in alleles or genome variations that can affect the makeup of that population's gene pool. Both emigration and immigration have connections to genetic drift (listed next), especially when an emigrating group colonizes a new habitat establishing a small "founder population". (Chapter 9)

9. Genetic Drift: The second major recognized and confirmed mechanism of genomic change, and one that has now been well documented by both logic and by countless genomic comparisons that show fixation or elimination of neutral or near neutral alleles and even some chromosomal mutations. Genetic drift is an especially potent force in small populations and again affects mainly neutral or nearly neutral genetic variations. (Chapter 9)

10. Neutral Evolution: Once viewed with suspicion by evolutionists overly fervent in the belief that natural selection ruled supreme in the arena of evolution, neutral evolution is now recognized as a very potent force in genomic change—arguably having as much (or more) effect on genomes as natural selection. (Chapter 8)

11. Symbiosis: From the complete endosymbiosis that led to mitochondria and chloroplasts in eukaryotes, to the more recent cases in aphids, hydrothermal vent clams, and others covered in Chapter 12, to the more common examples of mutualisms and parasitisms where at least significant coevolution has occurred in the involved symbionts, symbiosis has had profound effects on the evolution of most species—with many of the cases and details undoubtedly still unknown.

12. Long Spans of Time: The earlier argument against Darwin's explanation of common descent was that there was not enough time to allow for the divergence and specialization of all the Earth's biodiversity by natural selection and common descent. Today we know, but are

actually still unable to comprehend, the vast stretches of time available to evolution on this planet—surely sufficient to allow for the observed changes we presently observe, both in extant species and in the fossil record.

13. Physical or Geographical Separation of Intraspecific Populations: With allopatric speciation still likely accounting for a very large percentage of cladistic speciations over evolutionary time, these separations by various mechanisms (Chapter 13) have contributed greatly to biodiversity and common descent.

14. Extinction: In the broad sense of the term evolution means simply change, and extinction does very obviously change the makeup of biodiversity through time. It also often opens up opportunities in "ecological space" which other species may take advantage of—as mammals certainly did after the extinction of most of the dinosaurs (not counting the birds). Thus it "allows" evolution in some species along pathways that might otherwise be blocked.

15. Horizontal Gene Transfer: As discussed and argued in Chapter 7, these HGT processes have been an ever-present and ongoing set of mechanisms that have moved genetic and other genomic variations between individuals and species significantly—especially so in the Bacteria and Archaea. HGT has also resulted in the transfer of genes from endosymbionts into the nucleus of all eukaryotic organisms, and more cases are now coming to light showing HGT moving genes from microbes into eukaryotes.

16. Transposable Elements: These amazing genomic entities surely started out as parasites, but have either mutated into inert stretches of DNA or become functional as either parts of genes or regulatory sequences. They are a unique source of mutations as well. We still do not know in detail what percentage of our transposable elements have found functionality within the genome, though surely many of them are just useless parasites. (Chapter 6)

17. Epigenetic Inheritance: Modifications to the genome that do not change the DNA code itself, but do change the expression of the genes can be considered an epigenetic effect. If such effects are passed vertically between generations, they are termed epigenetic inheritance. Often due to environmental factors such as diet, temperature, disease exposure, etc. these modifications, in the form of molecular tags and other mechanisms, may prove to have facilitating effects on evolutionary change. We are still in a learning curve on this interesting topic. (Chapter 5)

18. Biodiversity: The countless speciation events that continually build biodiversity create a diverse biotic environment that provides numerous habitats and opportunities for other species to adapt to. All organisms provide hosts for parasites to adapt to and parasitize. When the first sizable trees evolved they created a habitat suitable for the evolution of epiphytes such as the bromeliads, climbing vines, most orchids, assorted lichens, etc. Before coral reefs, the many species adapted to live in coral reef environments could not have appeared. In short, biodiversity typically generates more biodiversity as speciation and adaptation specializes countless species to coexist in a variety of ways with the diversity of other life forms.

OTHER RECOMMENDED BOOKS ON EVOLUTION: WITH PERSONAL NOTES

1. ***On the Origin of Species*** by Charles Darwin. First printed in 1859 but reprinted countless times by a variety of publishers. The oldest book on my list, but still well worth reading to get a sense of how complete Darwin's thinking was at the time he wrote it. Of course we know far more today about several aspects and mechanisms that Darwin had no knowledge of, but Darwin is still "the giant" on whose shoulders all subsequent evolutionary biologists stand. He did after all discover natural selection, sexual selection, explained the origin of adaptations, and conceived the grand idea of common descent—among other contributions.

2. ***The Selfish Gene*** (new edition) by Richard Dawkins, (1989). Oxford University Press. A revolutionary book based largely on the ideas of William D. Hamilton concerning the centrality of genes in the evolutionary process. Dawkins clearly explains several deep theoretical ideas and argues forcefully about their importance in evolution and nature. He also adds a chapter on cultural evolution in which he coins the term "meme" for a unit of cultural evolution (to rhyme with gene), and this too is a very interesting essay. One will better understand much of nature after reading this book.

3. ***The Evolution of the Genome*** edited by T. Ryan Gregory, (2005). Elsevier Academic Press. This multi-authored work delves deeply into the many varied aspects of genome evolution including genome size, transposable elements, chromosome mutations, polyploidy, and many other interesting topics. It also includes chapters on comparative genomics that include

topics ranging from the specific to broadly general—simply a vast range of topics on an important sphere of evolutionary biology.

4. *What Evolution Is* by Ernst Mayr, (2001). Basic Books. I have read several of Mayr's books and consider him and William D. Hamilton to be the two greatest evolutionary biologists of the last century. This brief book is well written and easy to read. Yes, it is a bit dated now, but most of it is still important and fundamental to an understanding of evolution.

5. *This View of Life* by George Gaylord Simpson, (1963). Harcourt, Brace & World, Inc. Most of this book was actually published earlier, beginning in 1947, but this edition contains some more recently added chapters. Simpson has been called the father of American paleontology. He published over 1,000 scientific papers in his lifetime and wrote several books. This book is a gem of rationalist thinking and writing. It gives a truly broad overview of evolutionary and scientific thinking that is still perfectly relevant today. Simpson was a great writer, and I have gleaned many quotes from this excellent book. Of course it is dated and incomplete, but still worthy of being read by anyone interested in evolution or science in general.

6. *Evolution: What the Fossils Say and Why it Matters* by Donald R. Prothero, (2007). Columbia University Press. This excellent book is heavy on interesting evidence from the fossil record and contains sections explaining what science is (always a worthy topic) and why creationism is not scientific in any way.

7. *Tower of Babel* by Robert T. Pennock, (2002). The MIT Press. This is surely one of the finest of the multitude of books written about "the conflict" between science and religion in terms of evolution versus creationism. It includes a history of the conflict, philosophical issues associated with the topic, evidence for evolution, a great analogy using the evolution of language (thus the book's title), and more.

8. ***The Blind Watchmaker*** by Richard Dawkins, (1986). W.W. Norton & Company. Another very good book taking to task the fundamentalist creationism viewpoint. Dawkins again writes forcefully and convincingly as to the truth of evolution while pointing out the many problems and errors in the creationist view. Though untold numbers of books have addressed this controversy, this book and the ***Tower of Babel*** (already listed) are certainly two of the very best.

9. ***Full House*** by Stephen J. Gould, (1996). Three Rivers Press. Gould was the evolutionist other evolutionists loved to hate. He came across as arrogant, and he seemed to think he understood evolution better than any of his peers. I disagreed with him on a couple of points, but overall Gould did contribute much to evolutionary biology. He fought the good fight against creationist forces to keep non-scientific ideas out of the science classroom. In this book he takes up the mistaken idea (still prevalent) that evolution has been and is basically a progression from "lower forms to higher forms". It is a bit of a convoluted read containing a huge divergence into "baseball statistics", but it is still a good read, containing many good ideas and points about science and evolution.

10. ***The Logic of Chance*** by Eugene V. Koonin, (2012). Pearson Education, Inc. This is an awesome and interesting book, but also a challenging read that will not be understood by those without a good background in biology, genetics, and evolution. Evolutionary genomics is the primary topic here, but other topics such as symbiosis, the origin of life, the role of viruses in evolution, neutral evolution, and other topics in the molecular realm are introduced effectively as well. I am still reading and rereading sections of this interesting book.

11. ***Natural Selections: Selfish Altruists, Honest Liars, and Other Realities of Evolution*** by David P. Barash, (2008). Bellevue Literary Press. Barash is a prolific writer who promotes the ideas of Richard Dawkins in

inventive ways. You can see from the title of this book that it will be a bit unusual—and it is. I found it to be an interesting and varied mix of evolutionary biology and philosophy. He weaves together many ideas and examples in a synthetic way that makes the book that much more engaging than it might have been in the hands of another writer.

Index

Note: Page numbers with "*f*" denote figures; "*t*" tables.

A

Adaptations
  definition, 23
  fitness, 27–29
  forelimbs, 24
  Giardia lamblia, 26–27
  horizontal gene transfer, 25
  kiwi bird, 26
  modification characteristics, 24
  organelle loss, 27
  pseudogene, 27
  Raffelzia arnoldii, 26
  tapeworms, 28
  trait loss, 27
  venom genes, 24
  virulence factors, 25
Allopatric speciation, 112
  caves, 114–115
  continental drift, 113
  factors, 115–116
  geologic time, 114
  inoculation pods, 112–113
  islands, 112–113
  lava flows, 115
  ocean levels, 113
  rivers, 115
  tortoises and iguanas, 116
  uplift events, 113–114
Allopolyploidy, 120–121
Alu retroelement, 57–58
Ambystoma mexicanum, 97
Anagenesis, 4
Anomalocaris, 147, 147f
Autopolyploidy, 119–120

B

Bacterial conjugation
  cytoplasmic bridge, 62–63, 63f
  pilus, 62–63
  plasmid, 62–64
Balancing selection, 19–20
Burgess Shale fauna, 147

C

Chengjiang fauna, 147f
Choanoflagellates, 146
Chromosomal fission, 48
Chromosomal mutations
  deletion, 45–46
  duplication, 46
  fusions and fissions, 46
  inversion, 46
  neutral evolution, 76–78
  polyploidy, 48
  translocation, 46
  types, 47f
Cladogenesis, 120–121
Cladograms, 166–168, 167f
Co-dominance, allele combinations/
    multiple alleles, 41–42
Competition
  antibiotic-producing microbes,
    33–34
  black walnut tree, 31–32
  cooperation, 35–37
  corals, 34
  definition, 31–32
  insect species, 34
  interspecific completion, 36

Competition (*Continued*)
mutualisms, 35–36
offspring, 32
siblicide phenomenon, 33
survival of the fittest, 32
ubiquitous competition, 35–36
zooxanthallae, 36–37
Contingency
bird fluke, 155
continental drift, 156
dinosaurs, 155
ecosystem parameters, 154
exaptation, 152
land iguanas, 154
lateral fins, 152–153
meteorology, 154–155
natural selection, 151–152
Cyanobacterial endosymbiont, 102

D
Developmental processes
axolotl, 97
evolutionary diversification, 94
eye development, 98
genomes, 94–95
homeotic genes, 95–96
HOX genes, 96–97
locomotor mechanisms, 98
MADS genes, 95–96
teeth genes, 97–98
transcription factor, 95–96
Dihybrid cross, 40–41, 40f
Directional selection, 17
Disruptive selection, 18–19

E
Ediacaran fauna, 146
Edited mRNA, 74
Endogenous antifreeze proteins, 66
Environmental change
abiotic factors, 89
amoeba, 86–87
birds, 86–87
cave species, 88–89
components, 89

directional selection, 89–90
horseshoe crabs, 86
inter/intraspecific factors, 89
nautilus, 86
opportunities, 88
peppered moth, 89–90
suffered extinction, 87
whales, 86–87
Epigenetic mechanisms
acetylation and methylation, 49–50
advantages, 52–53
definition, 49–50
genomic imprinting, 52
inheritance, 51–52
liver and skin cells, 49–50
tags, 49–50
toadflax, 51
Exaptation, 24

F
Fossil record
bipedal theropods, 147–148
Burgess Shale fauna, 147
Chengjiang fauna, 147, 147f
choanoflagellates, 146
dinosaurs, 147–148
Ediacaran fauna, 146
Eusthenopteron and *Panderichthys*, 148–149
limb transition, 149
sponges, 146
transitional fossils, 149–150

G
Garbage in-garbage out principle, 44–45
Gene alleles, 40
Gene pool, 39
Genetic drift
bottleneck effect, 81
codons reading, 83
cytochrome C, 84
definition, 79
founder population, 80
Hardy–Weinberg law, 84
neutral base substitutions, 83

parent population, 80–81
red alleles, 82–83
sampling effect, 80–81
white alleles, 82–83
Genomes
architecture, 3–4
introns, 2–3
polyploidy, 119–120
transposable elements. *See also*
Transposable elements.
Genomic imprinting, 52
Giardia lamblia, 26–27

H

Hardy–Weinberg law, 84
HGT. *See* Horizontal gene transfer (HGT)
Homology
actin protein, 130–131
anagenesis, 129–130
bats and birds, 134
behaviors, 131
cladogenesis, 129–130
convergences, 133
feathers, 130
femur bone, 130
genetic code, 130–131
monkeys, 131–132
Pentastomida, 136
prehensile tails, 131–132, 133f
seahorses, 131–132, 133f
thylacine, 133–135, 134f
vampire bats, 135
Horizontal gene transfer (HGT)
adaptations, 25
conjugation
cytoplasmic bridge, 62–63, 63f
pilus, 62–63
plasmid, 62–64
definition, 61–62
endogenous antifreeze proteins, 66
endosymbiosis, 66–67
evolution, 65
intracellular digestion, 65–66
mergers, 65
transduction, 64

transformation, 64
HOX genes, 96–97

I

Imperfection
CD players, 138
femur development, 140–141
human genomes, 142
immune system, 141–142
land snails and slugs, 139–140
natural selection, 137–138
nostrils, 139–140
sexual selection, 141
spiders and insects, 139–140
swallowing problems, 139
vocalizations, 140
Inheritance, 39–40

J

Juglans nigra, 31–32
Jumping genes, 56–57
Junk DNA, 3–4

L

Lateral gene transfer. *See* Horizontal gene transfer (HGT)

M

Macroevolutionary change
axolotl, 127
definitions, 124
endosymbiosis, 125
evolutionary change, 123
feather acquisition, 124–125
horizontal gene transfer, 126–127.
See also Horizontal gene transfer (HGT).
parasites, 127
polyploidy, 125–126
self-fertilization, 126
speciation, 124. *See also* Speciation.
MADS genes, 95–96
Microevolutionary change, 123–124
Monohybrid cross, 40

Mutations
  asexual reproduction, 42–43
  chromosome. *See* Chromosomal
      mutations
  epigenetics. *See* Epigenetics
  gene regulation, 43
  noncoding DNA, 43
  point mutations. *See* Point mutations
  from transposable elements, 48–49
Myllokunmingia, 147, 148f

N

Natural selection
  adaptation. *See* Adaptations
  balancing selection, 19–20
  competition, 11–12
  definition, 10
  dinosaurs, 14–15
  directional selection, 17
  disease and parasites, 12–13
  disruptive selection, 18–19
  extreme changes, weather/
      climate, 12
  facts of nature, 9–10
  fire, 13
  living demands, 13
  misconception, 16–17
  pine tree, 14
  predation, 11
  sexual dimorphism, 18–19
  siblicide, 15
  stabilizing selection, 17
  superorganisms, 21
  survival of the fittest, 10
  unit of selection, 21
  water chemistry changes, 13
  weeding out, 10
Neutral evolution
  gel electrophoresis, 69–72
  genomic change, 69
  mutations
    chromosomal, 76–78
    in introns, 74
    neutral nonsilent, 72–74

  in noncoding DNA, 75–76
  redundant genetic code
    bases, 70–71
    codons, 70–71, 71f
    neutral silent mutation, 71–72
Noncoding DNA, 75–76
Nonsilent mutations
  amino acid substitutions, 72
  cytochrome C, 73
  earlobes, 73–74

O

Opportunity
  biodiversity, 163–164
  carnivores, 161–162
  chromosomal mutations, 163
  commensal, 160
  cysts, 160
  dinosaurs, 162
  ecosystem, 161
  evolution, 159–160
  humans, 164
  invasive species, 161–162
  parasitism, 160–163
  symbiosis, 160
  whales, 162

P

Parapatric speciation, 118–119
Photosynthetic symbiont, 108
Phyletic evolution, 4
Phylogeny
  advantages, 170
  definition, 4
  Animal Kingdom, 166–167
  cladogram, 166–168, 167f
  common/rare rule, 170
  Cycliophora, 166
  definition, 166–167
  fossil record, 166
  plesiomorphic and apomorphic,
      168–169
  polarity, 168–170
  primitive and lower characters, 168

rules of thumb, 169–170
tree of relationships, 165–166
Point mutations
    additions, 44–45, 45f
    arginine, 44
    deletions, 44–45, 45f
    garbage in-garbage out, 44–45
    single nucleotides, 43–44
    substitutions, 44
    transposable elements, 58
Polyploidy, 48, 119–120
Primary gene, 41–42
Progress
    adaptations, 175
    definition, 171–172
    mammal *vs.* reptile, 173
    trilobites, 174
    values, 171–172
Pseudogenes, 75–76

R

Raffelzia arnoldii, 26
Retrotransposons, 56–57
Rhagoletis pomonella, 116–117

S

Sickle cell allele, 19–20
Speciation. *See also specific speciation*
    without cladogenesis
        endosymbiosis, 120
        fusion speciation, 120, 121f
        hybridization, 120–121, 121f
    by polyploidy
        2N gamete, 119–120
        Down syndrome, 119
        sexual reproduction, 119–120
Stabilizing selection, 17
Survival of the fittest, 10
Symbiosis
    Arabidopsis, 103
    Buchnera, 106
    chemosynthetic bacteria, 105
    cilia and flagella, 103–104
    clams, 105–106

cyanobacterial endosymbiont, 102
engulfer cell, 104
genetic chimeras, 104
lichens, 108
parasites, 110
plant sap and aphids, 106
primary endosymbiotic events,
        103–104
role of, 101–102
sap-feeding insects, 106–107
secondary endosymbiosis, 104
symbiotic relationship, 109
Tetraselmis, 108–109
tsetse flies, 106–107
Sympatric speciation
    cichlids, 117
    disruptive selection, 118
    hawthorn/apple flies, 116–117

T

Tc-1/mariner, 57
Toadflax, 51
Tragopogon dubius, 120–121
Tragopogon mirus, 120–121
Tragopogon porrifolius, 120–121
Transduction, 64
Transformation, 64
Transposable elements
    jumping genes, 56–57
    junk DNA, 59–60
    mutation, 48–49, 58
    noncoding DNA, 58
    retroelements, 56–58
    retroviruses, 59–60
    selfish DNA, 58–59
    transposons, 56–57
Transposons, 56–57
Tyrannosaurus rex, 147–148

V

Viruses, 55–56

W

White lizards, 118–119

Printed and bound by CPI Group (UK) Ltd, Croydon, CR0 4YY

03/10/2024

01040426-0017